Drystone Walls of the Aran Islands

MARY LAHEEN is a practising architect and teaches at University College Dublin, where she graduated in 1981. She has a particular interest in cultural landscape, and considers that of Ireland an important but fragile and poorly protected part of Irish heritage. Associated with the Aran Islands for many years, she designed the Irish language school, Coláiste Ó Direáin, in the early 1990s. Other interests include Zen practice and Spanish language and culture.

Drystone Walls of the Aran Islands

EXPLORING THE CULTURAL LANDSCAPE

MARY LAHEEN

The Collins Press

First published in 2010 by
The Collins Press
West Link Park
Doughcloyne
Wilton
Cork

British Library Cataloguing in Publication Data
    Laheen, Mary.
    Drystone walls of the Aran Islands : exploring the cultural landscape.
    1. Cultural landscapes–Ireland–Aran Islands. 2. Land
    settlement patterns–Ireland–Aran Islands.
    3. Agricultural systems–Ireland–Aran Islands.
    I. Title
    333.7'6'0941748-dc22

ISBN-13: 9781848890251

Design and typesetting by Burns Design
Typeset in Goudy and Bliss
Printed in Malta by Gutenberg Press Limited

CONTENTS

Large portions of this book had their beginning, a number of years ago, as part of a master's thesis in Urban and Building Conservation at University College Dublin. Although somewhat unlikely as subject matter for an investigation into urban or building phenomena it was welcomed and encouraged by Loughlin Kealy, then Professor of Architecture and head of the Conservation graduate programme. I am most grateful for his broadminded approach, which allowed the pursuit of this subject which was not strictly within the parameters of conservation as it tends to be viewed in Ireland.

Since that time considerable changes have occurred in the Irish landscape, many of them detrimental to its heritage value and ecological balance. In parallel with these changes there have been developments on the global front, particularly among the conservation bodies such as ICOMOS, IUCN and UNESCO, which have been seeking to protect and raise awareness about the importance of landscape and the role of human beings within it. Hence, there is room for pessimism and optimism in equal measure, although it is difficult not to be alarmed at the extent of degradation of landscape that occurs on a daily basis.

The multidisciplinary approach that is required to understand a landscape has been a fascinating if at times perilous route of study. Being an amateur in the fields of other disciplines is like being a child again in a world of wonder and discovery. Among others Dr Matthew Parkes of the Geological Survey of Ireland was most helpful with my research into geological matters and I would like to thank him for his assistance. I am grateful to Dr John Feehan who read the thesis manuscript at an early stage and made helpful comments.

Undoubtedly, some of the seeds of curiosity and interest in Aran walls and constructions were sown in my mind during the many visits with students and colleagues from the School of Architecture to the Aran Islands, which used to occur on a yearly basis. These study trips, mostly with first year students, are now legendary, mostly for the mischief that students got up to on their visits west of the Shannon. However, the elemental nature of the structures on Aran, illuminated by the extraordinary quality of light there reflected from the ocean, offered a kind of touchstone for young people grappling with ideas about space, material and light and the merging of these in the design of buildings.

I would like to thank the people of the Aran Islands for their friendliness and willingness to answer numerous questions. I am especially grateful to

Dara Beag Ó Fatharta of Inis Meáin and to Tomás Ó Fatharta of Eochaill and his wife, Agnes. I am indebted to my brother, Máirtín Ó Laithín, for his assistance with conducting interviews, translation and other matters relating to the Irish language.

I am indebted to my friend, Djinn Gallagher, for an early proof-reading of the text and offering helpful advice.

Finally, I wish to thank Terry for his unfailing good humour and support, without which many things in my life would not have been finished on time or may not even have been contemplated.

For Patrick J. Laheen, 1922–2007

Introduction

FIGURE 1 View towards Ceann Gainimh, Inis Meáin.

While society in Ireland has been, for some decades, undergoing a process of urbanisation, as recently as the mid-twentieth century the country was primarily a rural and agricultural one. The possibility that the rural landscape of Ireland could be a repository of heritage initiated the exploration that has culminated in this book. The work was further propelled by the fact that we have lived through a time when that seemingly stable and continuing entity – the Irish countryside – has been subjected to irreversible and destructive forces of change.

The focus of this book is the cultural landscape formed by the drystone-wall field-boundary system of the Aran Islands. It is an agricultural landscape within a multilayered historic landscape set against spectacular natural beauty. Part of the catalyst for the investigation was curiosity about the forces governing the number and extent of field-boundary walls and their unique expression. The inquiry was also driven by a desire to understand the response of a community to the limitations of the land itself and to the difficult economic and political conditions which prevailed in the eighteenth and nineteenth centuries. A further reason for exploring the islands was the extent to which they form a 'built' landscape. Part of the investigation has been to figure out exactly to what degree the land or soil has actually been constructed. The Aran people have become legendary because of their ability to create soil from rock, although they were not the only ones who practised this method in the nineteenth century.

It became apparent early in the study that a land-division system prevailed across the islands, and investigation into this on a countrywide basis was the starting point of the research. Because of the manifest influence of geology in the cultural landscape, the methodology adopted was to investigate the geological structure of the islands. The primary source for this work was the 6" maps and memoir of the nineteenth-century Geological Survey of Ireland. Documentary evidence from the *Books of Survey and Distribution* regarding the fertility of the land and early (sixteenth- and seventeenth-century) references to the land-division system were sought out. It was possible to chart the changes that occurred in the field boundaries in the second half of the nineteenth century by comparison between the first edition of the Ordnance Survey (1839) and the revised survey (1899). In addition, the 1839 Ordnance Survey showed fields considered fertile or arable, and rocky pasture. This made it possible to establish an approximate pattern of land reclamation over time. Although the Ordnance Survey has not revised the 1899 map, its aerial photography carried out in 2000 gives a clear picture of changes that have occurred. Interestingly, there has been little change to the field boundaries in the 100-year interval.

Unfortunately, access to the papers of the Digby Estate (landlords of Aran from the mid-eighteenth century until 1922) has not been possible; although the papers exist and are held in the National Archive, the Archive is unable to locate them. Useful information regarding tenancies in the second half of the eighteenth century might be gleaned from the papers. We have little information for this period. Additionally, some light might be thrown on the riddle of the division of the fertile land, which occurs with such intensity in certain areas and can be seen by comparing the first and second editions of the Ordnance Survey.

The nineteenth-century interest in the Aran Islands among Celticists, archaeologists, Irish-language enthusiasts, artists and others meant that there was a body of work to explore regarding the islands, some of which had relevance to the study in question. The islands have also produced artists of their own, such as the short-story writer Liam O'Flaherty and his brother Tom, and the Irish-language poet Máirtín Ó Direáin. Their stories and poetry give glimpses of the life and work of the islanders. In more recent years, further work on the islands has been undertaken with respect to the geology, fauna and flora, traditions and customs of the islanders and history of the islands, which have also provided valuable sources. Tim Robinson, the writer and cartographer who lived on Inishmore in recent decades and who has written in detail about it, has provided a valuable resource. In particular, the fact that he established the land-division boundaries of the *ceathrúna*, and recorded them on his maps, has made possible the work on the land-division system in this study.

Although there are numerous fascinating layers of the cultural landscape in Aran, this study confines itself to the drystone-wall field-boundary system and the stone artefacts associated with it. In this regard, an understanding of the agriculture has been necessary. Fieldwork and discussions with farmers on the islands have been an important part of the study, as well as discussions with Teagasc, the research wing of the Department of Agriculture.

The first chapter looks at the concept of cultural landscape and its importance in the Irish context; the importance of interdisciplinary research in order to achieve a 'deep time' perspective on cultural landscape; and the context of European and worldwide guidelines on the management of cultural landscape.

Chapter 2 gives the reader a background history of the islands and the factors that have influenced the layer of the cultural landscape in question, such as topography, geology and political history. It also seeks to establish the historical and cultural importance of the system of land division, which on the large scale has influenced farming and settlement organisation. The land-division system has been researched in historical and geographical sources that apply to the country as a whole, or in some cases to Connaught, the province in which the islands are located. This research has then been applied to the Aran landscape as it is today, and as it appears in documentary and cartographic evidence.

Chapter 3 documents the settlement pattern, the intensification of enclosure that occurred during the nineteenth century, and the pattern of landholding as it now exists. The research draws on the history of Irish farming generally, and makes a comparative study of the first and revised editions of the Ordnance Survey. Other important sources such as the *Books of Survey and Distribution* and historical documentary evidence are used to speculate on how the enclosures and built heritage of farms evolved as they did.

Chapter 4 looks at one *ceathrú*, and a specific farm in that *ceathrú*, and describes the pattern of landholding and farming as it is practised today in the traditional methods of the islands. Photographic record and description document the elements of the built heritage that make up the agricultural landscape.

Chapter 5 assesses the value of this layer of the cultural landscape, and traces the changes in European and Irish farming which present a threat to traditional farming methods and consequently to the agricultural landscape. The development and progress of the Rural Environmental Protection Scheme is documented, as are the particular effects of this scheme on the Aran Islands. Proposals are put forward for supplementary measures that could make the conservation of this cultural landscape an exemplary one. The international charters are consulted as a guide to assessment and policy proposals.

Irish words are translated the first time they appear. Place names are generally given in the Irish spelling but are not translated except where the meaning has a particular significance. Maps drawn to illustrate points, and photographs, are the work of the author unless otherwise stated.

CHAPTER 1

Cultural Landscape in Ireland

The exceptional nature of the Irish cultural landscape has been understood and written about eloquently in recent years, not least because of the realisation that powerful agents of change, driven mainly by market forces, threaten its destruction. It is a landscape which is largely the imprint of earlier communities living in close connection with the natural world and using the terrain for the production of food. Within that landscape, it is now recognised that there are few if any wholly natural landscapes; on the contrary, it reflects human interaction with the land, forming accretions that had by the twentieth century produced a country rich in diverse cultural landscapes. The geographer Frederick Aalen considers that this 'cultural landscape is our major and most productive creation; it is both an artefact based on foundations of geology and climate, and a narrative, layer upon layer of our history and nature's history intertwined'.[1]

This is due in part to the fact that the country has been inhabited and farmed for so long. The history of agriculture, particularly around the coasts of Ireland, reaches back into the early Neolithic period. The constancy and continuity of settlement and agriculture mean the landscape is replete with field monuments, forming a remarkable landscape legacy.[2] The predominance of agriculture in the past, and the concentration on pastoral farming – the rearing of cattle and other livestock – in Ireland has helped to protect heritage in the landscape; removal of land forms in other places has often occurred to facilitate tillage farming.

Other factors have contributed to the richness and continuity in the landscape, such as the peripheral location of Ireland in relation to the European land mass, which in recent centuries delayed the effects of development and change as they occurred on the European mainland. Equally, in earlier times, the Celtic hegemony that was broken by the Roman conquest of Europe continued to exist in Ireland for over a thousand years. This part of our cultural landscape engenders particular interest and relevance outside Ireland, because the institutions, customs, language, settlement and field patterns, and husbandry reflect a society that was once common to the whole of Celtic Europe.[3] Until the final eclipse of the Gaelic order in the early seventeenth century, large parts of Ireland, particularly in the north and west, continued to be administered in the manner of the rural multi-kingdom that it had formerly been. Consequently, symbols of power and of society were invested in the landscape rather than in towns and cities as happened in other parts of Europe where varying factors, including the force of the Roman conquest, left a legacy of urbanisation. However, despite being on the edge of Europe, Ireland was constantly influenced by European culture, particularly from the medieval period onward. We continue to share ecological and cultural parallels, in particular, with those parts of Europe that are influenced by the great Atlantic from Galicia, Brittany, Cornwall and Wales to the Scottish highlands and Norway.[4]

We have already entered a time of great change. Agriculture is becoming less and less important to our economy, and the methods of agriculture itself have been subject to intense development in the last thirty years. As we shall see in later chapters, traditional agricultural practices no longer play a part in the real food-production economy of much of Europe. The agricultural landscapes that supported these practices are thus under threat. In addition, the rapid economic growth of recent years has precipitated huge infrastructural changes in road building and urbanisation which has had a destructive effect on cultural landscape in Ireland. This poses some questions: what will become of the cultural landscapes of Ireland if the activity that created and supported them – traditional agriculture – is no longer practised? Will the heritage and history that are contained and expressed in those landscapes be lost for future generations? And, if the cultural landscape of Ireland is an important part of the national heritage, how do we conserve it? Is it possible to conserve it and still maintain its authenticity?

In Ireland, rural landscapes of extraordinary natural beauty have certain legislative protections outside the realm of the conservation of the built environment. Likewise, field monuments and any structures built before 1700 are protected. However, the landscape that forms the setting of these monuments in most cases is not protected. Many consider the most tragic case of interference with this type of setting to be that of the extended landscape of the Hill of Tara in County Meath, a monument of great importance nationally and internationally. At the time of writing, this extended landscape is destined to be ripped apart to provide a motorway. While this is a high-profile case that has received a lot of attention, although not enough to prevent its destruction, there are many other cases of cultural landscapes in Ireland that have met their demise at the hand of the road builder. Although not as significant as Tara they are nonetheless important repositories of heritage of local or national value. International charters and legislative instruments exist which recommend policies for the management and protection of such landscapes, and would provide guidance and frameworks in these situations. However, neither the proposal by Meath County Development Plan (2007–2013) for the designation of the setting of Tara as a Landscape Conservation Area nor the fact that the Irish government was a signatory to the ratification of the European Landscape Convention saved this landscape from extensive infrastructural intrusion. The inclusion of Tara as one of the Royal Sites in the Draft Tentative List of proposed World Heritage sites, while too late to prevent the roadway and interchange being built, may serve to prevent further detrimental development to the setting.

FIGURE 2 A water trough for cattle in Árainn, beyond on the horizon is the cliff-top cashel of Dún Aonghasa.

Given the lack of appreciation demonstrated for such a significant landscape, what can we expect for the heritage that is embodied in the ordinary landscapes of Ireland? These agricultural landscapes – the ditches, hedges, boreens, outhouses, farm villages, settlement and land-use patterns – describe the habitation, cultivation and livelihoods of past generations. In their rich variety and distinctiveness, they reflect the social history and heritage of the country. They display, in general, the legacy of the small farmers of the last few centuries, although they also express other landscape forms and interaction, some of which are from the much more distant past. It is these landscapes that are currently being destroyed by forces over which we have less and less control. While the Irish cultural landscape is appreciated, understood and enjoyed to some extent by people, there is a general sense of powerlessness and apathy in the face of the landscape destruction which is occurring in Ireland at such a rapid and unprecedented rate. A recent report commissioned by The Heritage Council found that:

> recent European Environment Agency digital mapping of Europe's landscapes, [which] shows that Ireland has experienced unprecedented urbanisation and landscape fragmentation due to extensive new housing, major roads and other infrastructure projects. This has affected open countryside, villages and towns in all parts of the country, and the extent of the impacts is much greater than in other parts of Europe.[5]

Despite the fact that Ireland is a signatory to the European Landscape Convention (2002) – the first international treaty on landscape – there is a lack of leadership from government departments and statutory bodies, and apparently no clear policy on management and protection of cultural landscapes. In particular, there is in planning policy an absence of recognition that landscape is a repository of heritage. Article 5 of The European Landscape Convention obliges governments to:

a) recognise landscapes in law as an essential component of people's surroundings, an expression of the diversity of their shared cultural and natural heritage, and a foundation of their identity; and

b) establish and implement landscape policies aimed at landscape protection, management and planning through the adoption of the specific measures set out in Article 6.

The specific measures of Article 6 are: awareness raising; training and education; identification and assessment; landscape quality objectives; and implementation. While there is no doubt that the government is in the process of implementing these and other measures required by the European Landscape Treaty, it is not doing so at a pace sufficient to keep up with the penetrating and destructive forces facing the cultural landscape of Ireland at present. In this situation it is left to the communities who inhabit cultural landscapes and those who care about the landscape to take action to protect and safeguard what is left. Documents published by ICOMOS (International Council for Monuments and Sites), the

World Heritage Convention UNESCO and IUCN (International Union for Conservation of Nature), which offer guidelines for protection of landscape and record the experience of communities facing similar situations throughout the world, may be a useful starting point.

The term 'cultural landscape' was given an early definition by the American geographer Carl Sauer in an often-quoted passage from *The Morphology of Landscape* (1925):

> The cultural landscape is fashioned out of the natural landscape by a culture group. Culture is the agent, the natural area is the medium, the cultural landscape is the result . . . The natural landscape is of course of fundamental importance, for it supplies the materials out of which the cultural landscape is formed. The shaping force, however, lies in the culture itself.[6]

Much has been written about cultural landscape since then; however, the recognition that human beings shape the land as the land shapes them has prompted debate about the record of this interaction. The term began to be used by conservationists with the realisation that the intermingling of culture and nature is stored in the landscape, making of it a repository of heritage; a growing body of charters and legislative instruments relating to the protection of landscape in the international arena is testimony to the acceptance of this idea on a global scale.

The modern era of conservation of heritage is generally accepted as having begun at the Athens Conference of 1931. Located in the city of the ancient classical world, it dealt with the conservation of ancient monuments, (although an earlier manifesto was published by the Society for the Protection of Ancient Buildings in 1877). The Athens Charter was superseded by the International Charter for the Conservation and Restoration of Monuments and Sites (Venice 1964) published by ICOMOS in 1966, which outlined the basic principles that have since been accepted as fundamental to the conservation of buildings. It also upheld the importance of the setting of those monuments and buildings. Various other documents followed including the Florence Charter on Historic Gardens (1982) which provided definitions and guidelines in relation to designed landscapes. In 1979 at Burra, in South Australia, Australia ICOMOS adopted a new charter, using the Venice Charter as a foundation but in the very different arena of the Australian continent. The Burra Charter, guided by Australian conservationists, reflected the experience of conservation in a large country where European settlers had lived for 200 years, and Aboriginal people had lived for at least 60,000 years in profound connection with the natural environment physically, culturally and spiritually. This charter broadened the concept of patrimony and re-rendered the meanings of words such as 'place', 'cultural significance' 'conservation' and others to give a re-definition appropriate to the context. The charter was revised in 1988 and 1999 and the form used today is the Australia ICOMOS Burra Charter 1999, the others are archival documents. The Article 1 definition of the word 'place' included reference to site, area, land, landscape, and the explanatory notes encouraged a broad interpretation of the concept of place including, for example, trees, places of historical events, and spiritual and religious places. Linking the natural world with culture allowed for the emergence of the idea that a landscape could have cultural significance. It is now also accepted that, particularly in non-European cultures, landscapes with very little human interference can have important cultural or sacred significance; that landscapes can have intangible as well as tangible values; and that very often the intangible values associated with landscape are what give people a sense of identity.

In 1992 the World Heritage Convention UNESCO established the first international legal instrument to recognise and protect cultural landscapes. This reflected the broadening debate about the expression of cultural patrimony and a movement away from an Eurocentric perspective on heritage protection. Cultural landscapes were described as 'the combined works of nature and man' and would be inscribed using three categories:

i) landscape designed and created intentionally by man
ii) organically evolved landscape:
   a) a relict or fossil landscape is one in which an evolutionary process came to an end at some time in the past, either abruptly or over a period. Its significant distinguishing features are, however, still visible in material form;
   b) a continuing landscape is one which retains an active social role in contemporary society closely associated with the traditional way of life, and in which the evolutionary process is still in progress. At the same time it exhibits significant material evidence of its evolution over time.
iii) associative cultural landscape: the inclusion of such landscapes on the World Heritage List is justified by virtue of the powerful religious, artistic or cultural associations of the natural element rather than material cultural evidence, which may be insignificant or even absent.

In 1993 the spectacular volcanic region of Tongariro in New Zealand, which was already a World Heritage site for its natural qualities, became the first World Heritage cultural landscape (category iii) because of its cultural significance to the Maori. The following year the former World Heritage site Ayers Rock–Mt Olga in Australia was inscribed as the second World Heritage cultural landscape, now being called Uluru-Kata Tjuta and being managed by the Anangu, the traditional owners of the land. This represented years of work by the Anangu to confirm their role as custodians of their ancestral lands.[7]

The idea that intangible values associated with landscape could be an important part of patrimony had been considered for some time especially among indigenous groups, who sometimes had little or no built heritage as part of their culture. Among others, scholars who are associated with the Saami culture – indigenous people of northern Europe – advocated this idea.[8] Part of the Saami cultural ideal is that the landscape should be interfered with as little as possible; consequently, places of significance for them would have no human imprint. The Laponian area in the Swedish province of Lappland, traditionally inhabited by the Saami, was inscribed as a World Heritage site in 1996 for its cultural and natural values.

The landscape of Ireland, despite being largely a humanised landscape due to the continuing practice of agriculture for thousands of years, also held intangible values, which until recent years had great cultural significance. This is especially true of the Celtic landscape due to the non-urbanised nature of the society. The royal sites are a good example, as are the less well known places of outdoor assembly or parley sites of the dynastic families in the Gaelic world, which have recently been researched by Elizabeth Fitzpatrick.[9] They were still in frequent use at the time of the Tudor conquest. These are places in the landscape, often in remote countryside, where inauguration of the chieftain took place. They were also the venue for meetings of the *oireachtas*, an instrument of Gaelic government, for parleying and for gathering troops. Most of these sites are not built upon, although sometimes they are located at an ancient ring fort or ancient construction. They are generally located at a high place, usually within the territory of the hereditary inaugurator, with a commanding view of the landscape. There is some evidence to suggest that the *oireachtas* occurred on predetermined dates which coincided with the ancient Celtic festivals of *Bealtaine* (May) and *Samhain* (November). While there is often little or no physical evidence of heritage in such places, they had until the end of the sixteenth century a powerful political as well as cultural significance, and continue to be remembered in the locality. These places were not mapped after 1602 when they were no longer a threat to the Tudor government; however, they can be located through contemporary documentary evidence, often through the place name and through local tradition and folklore associated with them. Tara, of course was the most important place of outdoor assembly – symbol of the sacral kingdom of Ireland – it had tremendous political and cultural significance of both tangible and intangible value throughout history. That significance is still resonant today, albeit of a cultural nature, in contemporary Ireland.

The ancient system of land division in Ireland, which is explored in later chapters – a virtually invisible net of bounded and named units spread across the entire landscape – is another example of important intangible cultural significance which is not currently addressed in our consideration of landscape heritage.

In the Preamble to the Burra Charter the question of why we should conserve is answered as follows:

> Places of cultural significance enrich people's lives, often providing a deep and inspirational sense of connection to community and landscape, to the past and to lived experiences. They are historical records, that are important as tangible expressions of Australian identity and experience. Places of cultural significance reflect the diversity of our communities, telling us about who we are and the past that has formed us and the Australian landscape. They are irreplaceable and precious.

Another important aspect of the revised Burra Charter, taking into account the changes that occurred in conservation practice worldwide in the last decades of the twentieth century, has been the recognition that, people – especially those who are strongly associated with a place – should be fundamentally involved in the decision making, management and protection of a place. This principle has been accepted in the international conservation forum for some time. It is particularly relevant to the protection of cultural landscapes, because such landscapes are generally evolving entities where people continue to live and work. Ideally, the initiative to protect a cultural landscape should come from within the community who inhabit or belong to that landscape. In this instance the international guidelines and legislative instruments are simply an aid to develop strategies for management in order to achieve the protection and conservation of the environment that the community desires. Communities are sometimes suspicious of conservationists, fearing loss of control within their own environment. There have been cases of counterproductive conservation management; for example, in some of Africa's National Parks, indigenous nomadic people were excluded from their ancestral lands in the mistaken belief that their presence would be damaging for the natural environment. However, the folly of such an approach has been recognised and the principle of community involvement has been enshrined in more recent documents. In 2007 a fifth 'C' for Community was proposed by New Zealand to be added to the four existing 'Cs' – Credibility, Conservation, Capacity building and Communication of the

Operational Guidelines of the World Heritage Convention – in order to 'place humanity at the heart of conservation'.[10]

The Nara Document on Authenticity (1994) drawn up in Nara, Japan, further broadened the scope of conservation policy as it focused on the diversity of cultures within our human patrimony. It sought: to challenge current modes of thought within conservation practice internationally with particular regard to social and cultural diversity; to recognise and respect the cultural values of all societies including those which are marginalised or minorities often subject in our contemporary world to the forces of cultural homogenisation; and to uphold the principle of authenticity as fundamental to the clarification and illumination of our collective memory.

The Declaration of San Antonio, drawn up in San Antonio, Texas, in 1996, dealt with the Americas and with the pluri-cultural heritage of that continent. The autochthonous culture of the Americas had a rich cultural and spiritual relationship with the landscape, which comes down to us mainly in the form of intangible values. Although not completely destroyed by the European colonisation of the Conquest era, the heritage continues to be threatened by acculturation and by the homogenising forces of globalisation. The experience of conservation in such an arena provides useful guidelines and lessons which can be appropriate for strategies elsewhere.

In recent years, in parallel with ICOMOS and the World Heritage Convention (UNESCO), the International Union for the Conservation of Nature (IUCN) has turned its attention to the protection of cultural landscape. As a global environmental network, it has come from a background of nature conservation but has found in its efforts towards the protection of biodiversity that cultural diversity and biodiversity often coincide. Within the IUCN Protected Landscape system the focus for many years had been on natural landscapes of a pristine nature; more recently the idea that people play a part in the unfolding of nature through sustainable activity in relation to the land has been recognised. The emergence of the Category V Protected Landscape (IUCN) is testimony to the recognition that humanised landscapes, where people continue to live and work, often practising sustainable traditional agriculture, are important to our human patrimony, in both a natural and a cultural sense. The 'new paradigm'[11] also recognises the importance of people, specifically the people associated with the cultural landscape, both in terms of the role that they must have in management and stewardship of the landscape but also in terms of what can be learned from them with respect to sustainable practices in the care of landscape.

The World Heritage Convention has also made efforts to rectify what it sees as imbalances on the World Heritage List. Part of that strategy includes the understanding that our view of human patrimony must broaden to include the anthropological, multifunctional and universal as well as a purely architectural view of cultural heritage.

Hence we find that the conservation of cultural landscape is now an accepted idea, overshadowed for years by a focus on pristine natural landscapes on the one hand, and on the other by a view that heritage and culture reside in architectural monuments and by extension in cities and towns. While we have become accustomed to valuing a society in terms of its urban fabric, within the international conservation forum we have now reached a place from which the rural landscape can be viewed with equanimity and valued for its integral qualities, both natural and cultural. Similarly to the way that a medieval European city, for example, is venerated, the rural landscape can also record a way of life, be an expression of the spiritual and cultural beliefs of its people and can manifest the history of its society.

It should be remembered that conservation of landscape did not begin in the modern era but has been part of the age-old value systems of indigenous and traditional communities. The concepts of 'caring for country' and 'keeping country straight', an important part of the value system of the Aboriginal people of Australia, are an example of indigenous conservation practice. In Uluru-Kata Tjuta National Park some of the traditional ecological methods have now been adopted as part of the management practice for the park.[12] Care of the land was also valued in traditional farm management in Europe. Industrialisation of farming in recent times has undermined the role of farmers as the traditional custodians of the land. Part of the thrust for reform of the CAP (Common Agricultural Policy) has been to redress this. Curiously, in Ireland, before legislation was in place to protect, for example, field monuments, they were protected by folk beliefs. Damaging or interfering with a fairy fort or *ráth* would bring bad fortune; the Neolithic monuments at Brú na Bóinne, now a World Heritage site, were left untouched by local people for centuries in the belief that the fairies – powerful otherworld beings who could be malicious – dwelt within them.

FIGURE 3 Looking north towards Connemara over fields in Inis Meáin.

The Landscape itself as a Primary Source

Landscape is a complex entity. Thorough and complete study of landscape requires multidisciplinary work. This complexity is both its strength and its weakness; its strength because the richly diverse strands come together holistically in the landscape and reveal nature itself and humankind's complex relationship with it; its weakness because no single academic discipline can be its champion. Love of landscape itself must become the unifying factor. The lone explorer runs the risk of offending everyone in his or her specialist field, from the geologist and historical geographer to the farmer and tourist guide. It is, however, a risk worth taking for the sake of the future of the cultural landscape.

This book aims to explore a communal monument created by a population in the course of their daily work over many centuries; to investigate the heritage to be found in the rural landscape and to render it legible. Finally, it is an attempt to understand the cultural landscape in 'deep time',[13] that is, to perceive all the layers of the cultural landscape simultaneously. These layers include the geological, topographical, archaeological, historical, modern and ecological. Surprisingly, despite intense investigation of the Aran Islands over the last two centuries – from the prehistoric monuments to the blood type of the inhabitants – the agriculture landscape has not been explored. The significance of the other layers of cultural landscape on Aran is recognised, and, as we shall see in the final chapter, there are some protections in place for those, which by default also protect to some extent the field-boundary system.

FIGURE 4 Cattle grazing in summer pastures, Inis Meáin.

As a cultural landscape it is on the farthest end of the spectrum which calibrates the interaction between humankind and the earth. It is a humanised landscape that has an almost urban intensity; unlike those significant cultural landscapes which are replete with intangible significance but untouched by the imprint of human beings, the surface of Aran is almost entirely built or made by man; it is virtually a handmade landscape.

The urban comparison is suggested in part by the intensity and rectangularity of the built fabric, clearly delineated on the nineteenth-century Ordnance Survey maps, which show all the field divisions of even the smallest fields. The map is so densely subdivided that it looks very much like what we expect from the map of a city. Making one's way through the labyrinthine fields and lanes, map in hand, is similar to discovering an unknown city. And, for anyone who has traced burgage plots on a city map, the experience of tracing the field-boundary walls of Aran echoes with familiarity.

When one arrives on one of the islands, the 'city', from the ground at eye level, cannot at first be seen – it looks like a natural landscape, because of the uniquely integrated relationship between construction and landscape. Here, the

stone wall sits on a natural limestone rock and the boundary between the two is almost indecipherable. The fields of grass and wild flowers were laboriously prepared by men, combining sand and seaweed from the shore, perhaps as recently as sixty years ago. Yet it looks like 'landscape'. If you arrive by air, however, the mark of the human hand is startlingly visible, and with great clarity you can see the field-boundary system tracing its pattern over the landscape.

Nonetheless, this view of the landscape and comparison with a city is discipline-specific. The geologist sees rock formations and stone types in a millennial time frame that defies the conceptual mind. The botanist sees rare flora in a nutrient-poor soil and an absence of trees. The archaeologist sees monuments from the prehistoric to the medieval. The historical geographer sees settlement patterns, and so on. Using the landscape itself as the primary source for study in this way opens a window not just on the past, but also on the mystery of the human condition. We stand at this moment of our short lives, to recall Aalen's definition, unravelling layer upon layer of the narrative that describes our lives and nature's life in this particular part of the earth's surface, not knowing how it began or how it will end.

FIGURE 5 A field in Corrúch, Árainn.

CHAPTER 2

The Aran Islands – Geology and Land Division

Topography and Geology

Background History

The Townland Matrix

The Townlands and *Ceathrúna* of Aran

FIGURE 6 Eochaill evening.

There are four harbours between Heaven and Earth where souls are cleansed, the Paradise of Adam from which came the human race, Rome, Aran, Jerusalem. No angel who ever came to Ireland to help Gael or Gall returned to Heaven without first visiting Aran, and if people understood how greatly the Lord loves Aran they would all come there to partake of its blessings.[1]

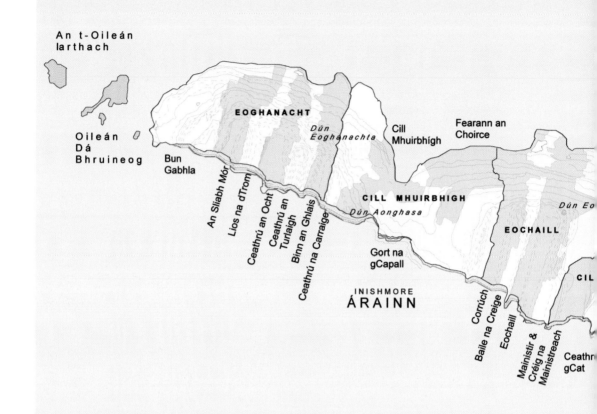

An t-Oileán
Iarthach

Oileán
Dá
Bhruineog

Bun
Gabhla

EOGHANACHT

Dún Éoghánachta

Cill
Mhuirbhígh

Fearann an
Choirce

CILL MHUIRBHIGH

Dún Aonghasa

Dún Eo

EOCHAILL

Gort na
gCapall

INISHMORE
ÁRAINN

An Sliabh Mór
Lios na dTrom
Ceathrú an Ocht
Ceathrú an Turlaigh
Binn an Ghlais
Ceathrú na Carraige

Corrúch
Baile na Creige
Eochaill
Mainistir &
Créig na
Mainistreach

CIL

Ceathr
gCat

Based on 1839 OS
Ceathrúna names & boundary lines from Oileáin Árann, a Map of the Aran Islands Co. Galway, Robinson T. (Folding Landsc
Terraces based on Geological Survey of Ireland 1867

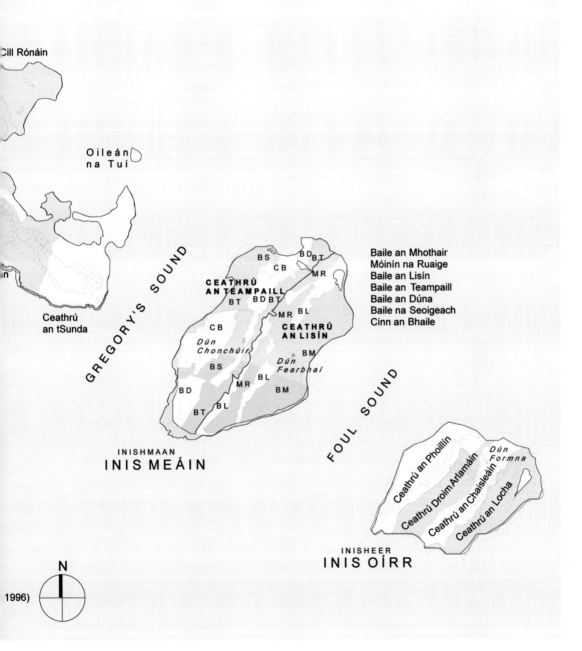

Cill Rónáin

Oileán
na Tuí

Ceathrú
an tSunda

GREGORY'S SOUND

BS   BD BT
        CB   MR
CEATHRÚ
AN TEAMPAILL
    BT   BD BT
              MR   BL
CB   CEATHRÚ
                AN LISÍN
Dún
Chonchúir   BM
    BS   Dún
              Fearbhaí
        MR BL
    BD   BM
    BT BL

Baile an Mhothair
Móinín na Ruaige
Baile an Lisín
Baile an Teampaill
Baile an Dúna
Baile na Seoigeach
Cinn an Bhaile

FOUL SOUND

INISHMAAN
INIS MEÁIN

Ceathrú an Phoillín   Dún
                          Formna
Ceathrú Droim Arlamáin
        Ceathrú an Chaisleáin
              Ceathrú an Locha

INISHEER
INIS OÍRR

N

1996)

FIGURE 7 Map of the Aran Islands.

Topography and Geology

The islands of Aran lie in a northwest-to-southeast direction and extend about 25 km in total. There are three main islands, known today in English as Inishmore, Inishmaan and Inisheer, and in Irish as Árainn, Inis Meáin and Inis Oírr. The islands have remained Irish-speaking despite the predominance of the English language in most parts of the rest of Ireland since the end of the nineteenth century. The largest island, Árainn, is more recently also called Inis Mór, although Árainn is the name that the older islanders use when describing their island in the Irish language. The name comes from the Irish word *ára*, which means 'kidney-shaped'. Árainn, the biggest island, being approximately 14 km at its longest, varies in width from 0.75 km to just over 3.25 km. Inis Meáin is smaller, approximately 5 km long and 2.5 km wide; the name means 'middle island'. Inis Oírr is the smallest; roughly circular, it has a diameter of about 4 km.[2] The name means 'eastern island' (it is east of the other two). There is also a number of much smaller islands: the Brannock Islands (Oileán Dá Bhruineog), and Rock Island (An tOileán Iarthach) off the western tip of Árainn, and Straw Island (Oileán na Tuí) at the edge of Cill Éinne Bay.

These islands, separated from the west coast of Ireland by about 8 to 9 km, are known to have been inhabited since 2500 BC but may have been settled as early as 4000 BC. Seemingly inhospitable, their location was, however, attractive for various human activities in the course of history. They may have been geographically central to an early Celtic seafaring empire while, some centuries later, their remoteness was attractive to the Early Christian monks who sought extreme conditions for prayer and retreat. By the time the Irish church, under the influence of the Orders and the centralising Roman church, had become part of a Europe-wide dissemination of Christianity, the islands were a popular centre of pilgrimage. To the Gaelic lords who controlled the port of Galway in medieval times the islands were strategically located. Yet in later centuries, they became peripheral again, and that remoteness from the centres of power during the centuries of colonisation and plantation allowed the survival of the Irish language as well as many aspects of the cultural landscape that were lost elsewhere in the centuries of development, industrialisation and change.

Equally relevant to the making of the cultural landscape is the story of the islands before the arrival of human beings, poetically described by a recent chronicler of Aran:

> At one point a few trodden footholds lead down from it to a recess almost under the *roidín* ('little road') itself, from which the hart's-tongue fern sticks out its long pointed fronds, and cool water lies in a small stone-lined basin

below a thick ledge of limestone. A few flattish stones lie at the rim of the basin like uncomfortable kneelers before a shrine; if you stoop and peer under the great limestone altarpiece you see the shale at the back of the recess. It glistens with moisture, and it comes away in horizontal flakes when you pick at it. It looks like the edge of a huge mouldering book, shut forever by the weight of rock above it, and in fact it is the history of one of those fleeting lands whose hills were weathered away and carried off as mud by rivers, and piled in layers on the sea floor during the gestation of Aran.[3]

That gestation occurred, geologists tell us, 325 to 350 million years ago when the nearly horizontal beds of limestone and shale were laid down in warm tropical seas during times of raising and lowering of sea levels.[4] These horizontal beds and the vertical fractures through the rock, called joints, which formed as a result of ancient earth movements, are the two primary geological formations of the islands and have greatly influenced the shape of the cultural landscape. Because of the paucity of soil cover over the rocks, the geology of Aran is more apparent than in other places. While walking around the islands one can see clearly the characteristic layering of the rock as the inland cliffs take great giant-steps down to the northern coast, laying out the shale terraces before them. At the southern Atlantic coast the geological formation can be seen with even greater clarity where the layers of limestone and shale are piled up on top of one another forming high cliffs, often more than 50 m high, and at the highest exceeding 120 m.[5] As the shale beds are softer than the limestone they are more easily eroded. On the cliff face where this erosion occurs, the limestone beds jut out over the edges forming great ledges of stone that are characteristic of the Aran cliffs. This predominant layering of the landscape echoes the escarpments of the nearby Burren on the mainland of County Clare, which is part of the same carboniferous limestone geology.

The shale is an impervious material and consequently plays a part in the hydrology of Aran. The ground water which seeps down through the vertical cracks in the limestone – the joints – is held by the impervious shale layer at the base of each limestone terrace until it emerges distilled at the ledges of the southwestern sea cliffs or, at the base of the inland cliffs in pools of fresh water.[6] The former is returned to the ocean and can be seen spilling in torrents from cracks in the rock at times of high rainfall or dripping from the crevices above the block beaches in dry weather. The water that emerges at the inland cliffs is destined to provide what in former times was a crucial requirement for settlement – fresh water for human consumption. Not surprisingly the villages of Aran occur in close proximity to these wells found at the base of the inland cliffs. Settlement occurs there not only because of the sources of fresh water but also because the shale terraces support fertile land and therefore the possibility of agriculture.

In general, the shale beds are thin but occasionally are quite thick, up to 1.5 m in places.[7] They lie at varying distances apart, but about 12 m is a common interval. The bedding planes of the limestone, interleaved with beds of shale, were originally horizontal but shifted due to pressure from earth movements about 290 million years ago, which tilted them slightly towards the south.[8] In this way the islands are a classic example of a *cuesta*, with the dip on a very gentle slope facing southwest, and the escarpment facing northeast. The dip is generally 3° to 5° and occasionally up to 10° to the southwest on Árainn and Inis Meáin. On Inis Oírr, the beds dip gently to the south or southeast at 1° to 3°.[9]

While the shale terraces lie beneath the fertile enclosed fields of the villages, Aran's spectacular diversity of limestone pavements also have a part to play in the yearly agricultural cycle. These pavements of pale-grey to blue-grey rock, rich in fossils, which are the remains of the plants and animals that inhabited the tropical seas during the millennia of the rock's formation, with scant grass but dry underfoot, provide excellent wintering for cattle. In common with the farmers of the Burren, the Aran community has made good use of this seemingly inhospitable terrain at the high terraces stretching toward the Atlantic cliffs.

The vertical joints that traverse the rock and allow water penetration to the shale terraces below appear as regular and parallel sets of fissures in the limestone pavement. These originally occurred due to pressure sustained by the rock during ancient earth movements but continue to be opened and enlarged by rainwater erosion. On Árainn and Inis Meáin, the main set of joints runs in a direction 10° east of north and a secondary set runs east–west, forming the high cliffs at the south of the islands. On Inis Oírr, the joints are N17°E and N12°E, occasionally crossed by others running E25°N for short distances.[10] The transverse valleys which occur on the islands, and the submerged valleys between the islands at Gregory's Sound and Foul Sound, have a similar bearing and have also developed along joints in the strata. However, the valley stretching towards the southwest at Cill Mhuirbhigh, known as the Blind Sound, and part of the valley at Cill Éinne are thought not to have developed in this way but are due to marine erosion at a time of higher sea levels.[11] It is these vertical joints in the rock, delineated as they are by the stone boundary walls built along them, which give the island landscape its somewhat unexpected orthogonal character. This is particularly true on Árainn where the rectangularity of the densely webbed fields have an almost urban orthogonality.

These two geological formations – the shale terraces and inland cliffs, and the sets of vertical parallel fissures running roughly north–south through the limestone beds – which combine to create the particular hydrology of Aran, have,

FIGURE 8 Erratic on the limestone pavement at Mainistir in Árainn.

as we shall see later, directly influenced both the farming and settlement pattern, and the field and land-unit-division boundaries of the islands.

While these formations were occurring in geological time, so long ago that human beings have difficulty conceptualising it, the erratics – boulders of granite, sandstone and limestone strewn intermittently across the landscape – are of much more recent origin. They are the remains of the glacial drift and boulder clay carried over from Connemara and the Galway coast by the receding ice at the end of the last Ice Age, about 10,000 to 15,000 years ago. Geologists think that all of the islands were covered by boulder clay immediately after the ice melted, but that much of the till was soon eroded by Atlantic gales, leaving mainly stones behind.[12] The large erratics remain in their original positions, now sometimes perched on pedestals that have been carved out of the limestone pavement by the continuing erosion of wind- and rain-driven particles. The lesser stones have provided the raw material for the cultural landscape of Aran, the imprint of humankind on the landscape since prehistoric times.

The most striking attribute of the cultural landscape today is the densely webbed pattern of fields enclosed by drystone walls. The fields themselves, many of them reclaimed from bare rock over the past three centuries, contain farming artefacts such as troughs, stiles, gaps, sheds and animal enclosures, all made from the ubiquitous stone. Also inhabiting these densely enclosed spaces are the stone monuments of earlier generations, including the wedge tombs of the Neolithic period, the circular drystone-wall cashels of the Celtic cattle-rearing people and the ruined stone churches and sacred monuments of the Early Christian period.

Background History

The date for settlement of 2500 BC or even as early as 4000 BC is garnered from the wedge-tomb monuments of Aran, of which archaeologists speculate there were possibly half a dozen in total on the islands.[13] At least one survives on Árainn, south of Corrúch village in the townland of Eochaill, and there are the remains perhaps of another at Fearann an Choirce. There is also a ruined wedge tomb at Ceathrú an Lisín on Inis Meáin. The wedge tombs are one type of funerary-stone monument of the Neolithic people – farmers who settled around the west coast where soil cover was thin and forest more easily cleared than in the densely-forested inland territory.[14] Archaeological evidence in recent years such as that found at the Céide Fields in north Mayo shows that Neolithic groups were involved in cattle-farming in fields enclosed by curvilinear and parallel stone-wall or ditch boundaries. It is also now known that those who lived near the coast also used the resources of the sea. In other words, the fisher-farmer settlement that one finds in coastal regions today appears to be similar to the model for Neolithic settlement.[15]

Chronologically, the next dominant layer of settlement evident in the islands is that of the stone forts, thought to have been built by cattle-rearing, and probably seafaring, Celtic people. Dún Aonghasa and Dún Dúchathair (classified in archaeological terms as promontory forts) on Árainn are considered the earlier ones – seventh century BC to first century AD – and the inland cashels of Dún Eoghanachta and Dún Eochla on Árainn, Dún Chonchúir and Dún Fearbhaí on Inis Meáin and Dún Formna on Inis Oírr are thought to be later – first to seventh century AD. These inland cashels are considered to be stone versions of the ring forts (also called raths) that proliferate in the rest of Ireland. They are the dominant settlement form of the Celtic society whose foundations were probably laid in the Bronze Age (2000 BC – 500 BC). Early records show that Ireland and Britain were already Celtic-speaking, and organised in the same way as related continental communities, as early as 500 BC. This Celtic hegemony was broken in Europe and Britain with the expansion of Rome, but continued to dominate in Ireland until the Middle Ages.[16] Raths seem to have been in use as dwellings in the north of Ireland as late as the seventeenth century.[17] However, some hold the view that ring forts, such as the O'Hagan homestead in Tullahogue drawn by military cartographer Richard Barthelet, were inhabited only in times of war.

Some speculate that the promontory forts, and in particular Dún Aonghasa (despite its appellation, *dún* meaning a king or lord's house in early Ireland, and Aonghus being the name of the legendary leader of the Fir Bolg) were not 'houses' or centres of commercial enterprises connected with the land or the sea but had a function to do with worship, sacred ritual and assembly.[18] If so, they were a precursor of the next wave of habitation in Aran, for the islands were to become an important centre of Irish monasticism in its eremitical Early Christian form from the sixth century to about the twelfth century. The spread of the Early Christian monasteries was accompanied by widespread economic revival. The monasteries themselves were self-sufficient and it is known that monastic communities engaged in farming and fishing.[19] There are records in early Irish manuscripts of various grants of lands to the Church from the Gaelic lords.[20] Legend records the granting of land in the islands of Aran to St Enda by Oengus, King of Cashel, member of the Eoghanacht sept.

Archaeological investigation has yet to be carried out, but it is known that ancient settlement and field walls exist under the sand at Teaglach Éinne (the church of St Enda) on Árainn between An Trá Mhór and Gregory's Sound. These were written about shortly before the Geological Survey in 1861 by the vicar of the time, the Rev. Kilbride, who was living on Árainn and interested in archaeology. The discovery was made when a large amount of sand was blown from the strand into Gregory's Sound, reducing the level of the ground by 3 m to 6 m and revealing the presence of field and garden walls as well as clochans. An account of the occurrence is quoted in the *Geological Survey Memoirs*.[21] Since that time, blown sand has covered all the remains again. It is assumed that these walls and buildings predate the Early Christian church of St Enda, which was built above the level of the settlement described.

By the later Middle Ages, ownership of the islands was attributed to the O'Briens of Munster. The Annals of Connaught record the death of 'Tadg O Briain, lord of Ara' in 1474. One hundred years later, ownership of the islands was disputed between the O'Flahertys of Connaught and the O'Briens. Under the Gaelic system, freeholders who were kinsmen or followers of the O'Flahertys or the O'Briens would have farmed the land of the islands.[22] Both clans were then outmanoeuvred by Queen Elizabeth I. Following the annexation of the islands into the queen's own dominion in the late sixteenth century, the islands passed through the hands of various owners, including Richard Butler, fifth son of the first Duke of Ormond, who became Earl of Arran in 1662.[23]

The islands were bought in total by the Digby family in the middle of the eighteenth century, and they and their descendants remained in possession, as absentee landlords, until the land was bought under the Land Commission scheme in 1922. In 1927, the Land Commission, now an organ of the new Irish state, granted the farms to the tenants, on payment of an annuity. These farms are generally in the ownership of those same families today.

The Townland Matrix

Unravelling the intricacies of the drystone-wall landscape of Aran requires an investigation into what geographers today call the townland matrix. In contemporary Ireland the townland matrix is the remarkably intact vestige of the ancient land-division system of the country. This territorial framework, with the *baile* (township) as the unit of land measurement, is known to have been in place by the twelfth century.[24] The historian Kenneth Nicholls speculates that many of the units may have been laid down as early as the seventh century, and some even before.[25] By the late medieval period, the whole country was composed of a net of these units, each of which had a name and a boundary, and was normally the holding of a sept or lineage. Often the name of the sept was ascribed to the land unit: *Baile Uí Dhubhda* (Ballydowd, County Mayo), *Baile Mic Aodhagáin* (Ballymacegan, County Tipperary),[26] but topographical names and those related to historical events are also common.

The *baile* was a sustainable land unit – 'a township, in the historical sense of a rural settlement – with all the necessary types of land for a self-contained economy'.[27] It included a small amount of fertile land and a larger amount of less fertile land. If there was mountain, woodland or bog in the vicinity it was held in common, each *baile* claiming the part that was nearest to it. Some time after the establishment of this network of units they were assessed and divided into fractions of the unit. The size of the fraction corresponded to their economic value in terms of livestock that could be fed or crops that could be produced, rather than a strict acreage. Accordingly, while there are sometimes large differences between the size of the assessment units, there is an inbuilt equivalence between them.[28] Some indications show that in Connaught, the assessment system based on the division of the *baile* into four *ceathrúna* (quarters) was created by the twelfth-century O'Connor king Toirdelbach Mór Ua Conchobair or his son Ruaidhrí.[29] Sometimes the township is subdivided into *trian* (thirds), anglicised as trine or treen. The quarter or trine was divided again into cartrons (a word of Norman origin[30]) and gneeves. There is a strong provincial unity in Connaught, including County Clare where one *baile* equals four quarters, sixteen cartrons, or twenty-four gneeves. There are some differences in the names of the assessment units and the sizes of them among the provinces. For example, in Ulster the *baile* is known as a *bailebetagh* and is considerably larger than the large unit in Connaught, but it is clear that an overall system was applied to the whole country.

This type of system is not unique to Ireland. There are parallels in the land division of pre-Norman England where land was divided into vills, trefs, townships, etc. What is remarkable is that the matrix, despite the vicissitudes of history, still exists in Ireland, although not in its complete and original form. It is known that in the east of the country, Anglo-Norman manors were laid down on the ancient territorial framework.[31] The parishes likewise followed the boundaries of what we now call townlands. At the time of the Munster Plantation (1583) the confiscated lands of the Earl of Desmond were measured for the purpose of granting to English settlers. However, the boundaries of the ploughlands of Munster remained the same despite the change of ownership, and notwithstanding maps prepared by Elizabethan surveyors which showed geometrically defined farms with orthogonal boundaries bearing no relation to the physical realities of the landscape.[32]

Thanks to the continuing efforts of the Elizabethans and those who came after them to establish the size and acreage of these ancient land units for purposes of taxation or confiscation, we are able to trace their existence. In the details of the Survey of Connaught carried out under Sir Thomas Wentworth, later Earl of Strafford, in 1636–1637, names and acreage are given for the townships and the number of fractional units are given; an estimate of the profitability and type of land, whether arable, rocky pasture, etc., is also given.[33] Unfortunately, the maps were lost but the details remain in the *Books of Survey and Distribution*, which record land being forfeited and granted in the Restoration land settlement of 1640–1641. The native land-division system also provided the framework for the mid-seventeenth century Down Survey (1654–1656) although Sir William Petty seemed exasperated by the plethora of names of the various land units he encountered in different parts of the country and by what he saw as their inequality: 'As to these townlands, ploughlands, colps, gneeves bullibos, bullibellas, horsemen's beds: they are at this day manifestly unequal both in quantity and value, being made on grounds that are all obsolete and antiquated.'[34]

Yet, as late as the census of 1821 the areas of what were now being called the townlands were still apparently unknown.[35] Part of the impetus and argument in favour of a survey of the whole country at the beginning of the nineteenth century was that the townlands had to be accurately surveyed for purposes of valuation.[36] What later became the first Ordnance Survey was initially referred to as 'the new townland survey'.[37] During the first Ordnance Survey, it was the fractional assessment units which were then recorded cartographically as townlands. Numerous townlands now existing in the west and northwest have the prefix 'Carrow' from the Gaelic *ceathrú*. Some new units were made but generally this was done by amalgamating existing townlands, or by subdivision, which did not damage the overall framework. Studies in Ulster show that it is possible to reconstruct the original *bailebetagh* from the modern townlands.[38]

The townland matrix reflected the spatial dimension of Gaelic society which was non-urban in nature. It was a countrywide landscape master plan that reflected an intimate understanding of the land and divided its resources into usable and sustainable parcels for the purpose of agriculture, the principle being that the equivalence between the units resided in the potentiality of the land and not in the quantity. This is something that may never have been fully understood by the administrators of the land in the centuries after the demise of the Gaelic order. Later in this chapter we will see how clearly that principle is manifested in the landscape of Aran, even today, both at the scale of the whole island and that of the individual farm.

The Townlands and *Ceathrúna* of Aran

In his discussion of the land-division system, archaeologist Thomas McErlean points out that, although the tenurial framework imposed on the landscape was largely present from the beginning of the Irish medieval period, with origins even further back, it should not be assumed that society was in any sense static: 'Population could wax and wane without altering the boundaries of the land units, but these could exert a strong influence in directing and controlling dynamic settlement.'[39] He also holds the view that the territorial organisation was conceived to manipulate and exploit the land in order to support the superstructure of society.[40] It is in the context of this fiscal aspect that we meet an early documentary reference to the land division in Aran.

Attached to each sustainable land unit was a due to be paid to the Gaelic overlord. This exaction took the form of, in some cases, a cow or other animal, an amount of butter or even lodgings for the lord and his retinue. Referring to these exactions we find a reference in *The Compossicion Booke of Conought* recorded in 1585. In the context of an indenture between the Irish lords of Iar Connaught – mainly the O'Flahertys – and Queen Elizabeth I, describing the lands of Iar Connaught, which were now to be levied with a fee of 10 shillings per quarter (of land) per year payable to the queen instead of the various exactions due from the land to the Irish lords, the islands are described as follows: 'In the Barrony of Aren there are three Ilands, one called Arenmore consisting of 24 quarters and the other two consisting of 6 quarters a peece, which in the whole cometh to 36 quarters being the queens Ma ties inheritance.'[41]

The quarter referred to is the English translation of *ceathrú*, the fractional unit of the larger land unit known as a *baile*.

In the *Books of Survey and Distribution*[42], the entry for Aran, taken from the 1636–1637 Strafford Survey describes 'the half-barony of the Isles of Aran': 'the great Island' having six quarters in 'Killeny,' six quarters in 'Oghall', six quarters in 'Kilmurry', and six quarters in 'Oghnought'. The middle island, called 'Inishmaine', has four quarters in 'Kilcannon' and four quarters in 'Liske'. The 'Small East Island called Inishsure' contains four quarters. Acreage and estimated good and arable pasture is given for each of them.

In 1684, Roderic O'Flaherty in his *Chorographical Description of West or h-Iar Connaught* describes the islands similarly:

Ara-Mhor, the greatest and furthest to the west of them, contains twenty-four quarters of land, and is twenty-four miles in compass; wherein on the southside, stands Dun Engus . . . The middle island of Aran contains 8 quarters of land . . . The third island of Aran, Inisoirthir, or the Eastern Isle so called of its situation from the two other, contains four quarters of land with a castel on a height.[43]

By 1839, at the time of the first Ordnance Survey, the four townlands of Árainn are recorded as Onaght, Kilmurvey, Oghil and Killeany. The names are the same, although the English spelling has changed slightly since 1636. The two townlands of Inis Meáin are Carrownlisheen and Carrowntemple. These names have changed since the Strafford Survey, when they were Liske and Kilcannon. Inis Oírr is one townland, as it was in 1636.

Comparing the size of the modern townlands of Aran to those in the rest of County Galway, the Aran ones are generally larger. This is because on the mainland, the fractional unit, the *ceathrú*, *trian* and sometimes *cartron*, seemed to become the modern townland. However, on Aran the original *baile* was retained as a modern townland and marked as such on the first Ordnance Survey map, but the *ceathrúna* boundaries were not recorded. Fortunately, the writer and cartographer Tim Robinson walked the boundaries with local people in the 1970s and ascertained the names and boundaries of the *ceathrúna*, most of which were still known, and subsequently recorded them on his map of the islands.[44]

Today in Árainn (Inishmore) there are four townlands: Eoghanacht; Cill Mhuirbhigh; Eochaill; and Cill Éinne. Eoghanacht is the most westerly townland and the name of the area may predate the *baile* boundary. Dún Eoghanachta, a stone cashel of early date, has a focal position in this townland. The antiquarian and Gaelic scholar John O'Donovan, writing at the time of the first Ordnance Survey, felt that the fort was not built by the Eoghanacht 'but no doubt they dwelt within it'.[45] The Eoghanacht Árann were one of the seven Eoghanachta, listed in the genealogical tracts, who ruled Munster until the tenth century (Aran was part of Munster until the Middle Ages). They ruled over Aran and the neighbouring part of the Burren, but 'disappeared from history at an early date'.[46] The King of Cashel – Oengus Nad Fraoich, mentioned earlier – who reputedly gave the islands of Aran to St Enda, was an Eoghanacht king. Six quarters exist in this townland today,[47] their boundaries and names are known: Bun Gabhla; An Sliabh Mór; Lios na dTrom (with grazing rights on Oileán Dá Bhruineog); Ceathrú an Ocht; Ceathrú an Turlaigh (plus An tOileán Iarthach); and Binn an Ghlais.

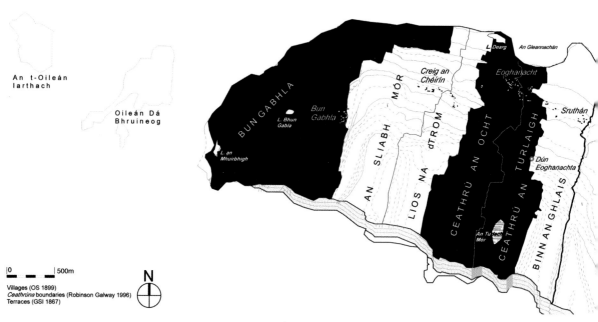

FIGURE 9 Map of the townland of Eoghanacht, Árainn, showing *ceathrú* and village relationship.

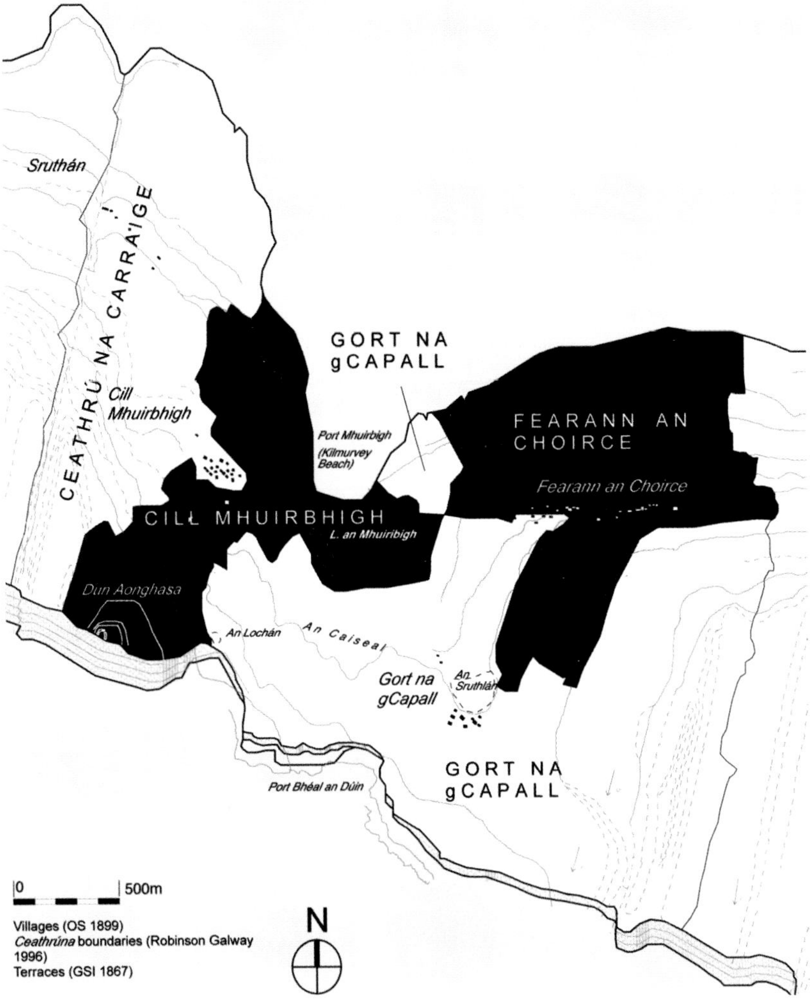
Sruthán
CEATHRÚ NA CARRAIGE
Cill Mhuirbhigh
GORT NA gCAPALL
Port Mhuirbhigh (Kilmurvey Beach)
FEARANN AN CHOIRCE
Fearann an Choirce
CILL MHUIRBHIGH
L. an Mhuirbhigh
Dún Aonghasa
An Lochán
An Caiseal
Gort na gCapall
An Sruthlán
GORT NA gCAPALL
Port Bhéal an Dúin
0
500m
Villages (OS 1899)
Ceathrúna boundaries (Robinson Galway 1996)
Terraces (GSI 1867)
N

FIGURE 10 Map of the townland of Cill Mhuirbhigh, Árainn, showing *ceathrú* and village relationship.

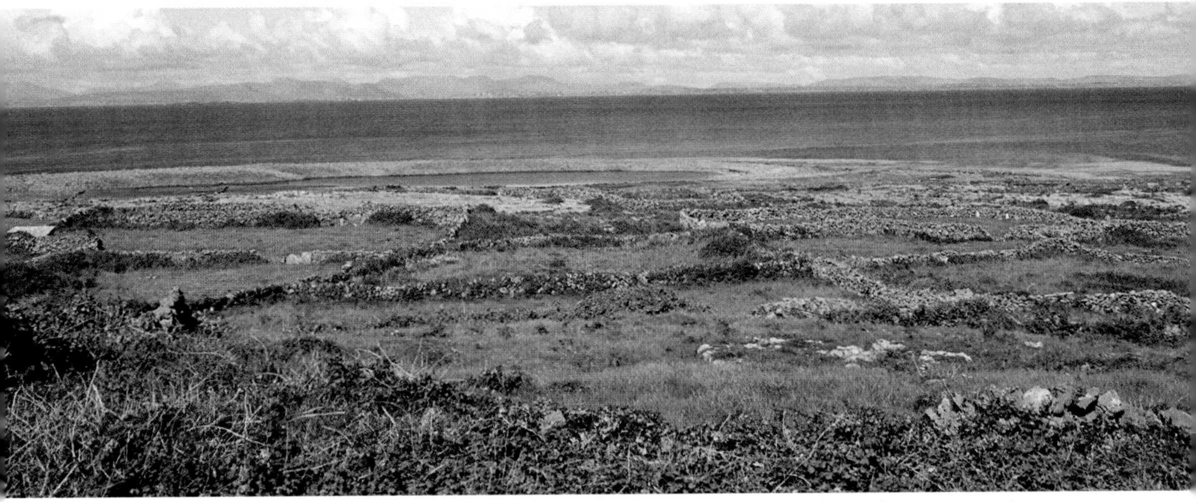

FIGURE 11 View over Ceathrú an Ocht and Lios na dTrom; the freshwater lake Loch Dearg is at the centre of the picture .

Moving eastward the next townland is Cill Mhuirbhigh meaning 'church of the sea plain', which may refer to Teampall Mac Duach, an eighth- or ninth-century church (opinion among archaeologists is divided: it may be as late as eleventh century) in use until at least the fifteenth century, located on the sandy flat land surrounding Port Mhuirbhigh. It is now in the quarter of the Kilmurvey House farm. Four of the six *ceathrúna* are known: Ceathrú na Carraige; Cill Mhuirbhigh; Gort na gCapall; Fearann an Choirce. Two *ceathrúna* exist within this townland but the boundaries and names are lost.[48] The important stone cashel or promontory fort known as Dún Aonghasa, is situated in this townland.

Eochaill is the townland in which the stone cashel of Dún Eochla is situated. Who built this cashel is not known – perhaps it was also an Eoghanacht cashel. There are four known *ceathrúna* boundaries here: Corrúch; Baile na Creige; Eochaill; and Mainistir. The boundaries of the *ceathrúna* of Mainistir and Eochaill are believed to contain two *ceathrúna* each.

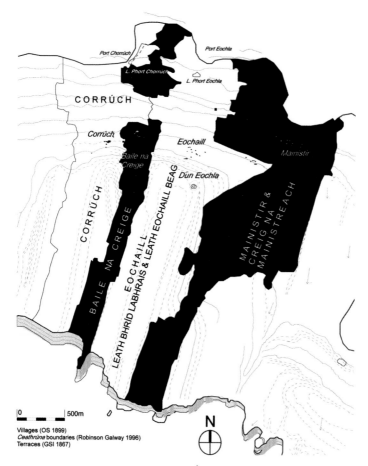

FIGURE 12 Map of townland of Eochaill, Árainn, showing *ceathrú* and village relationship.

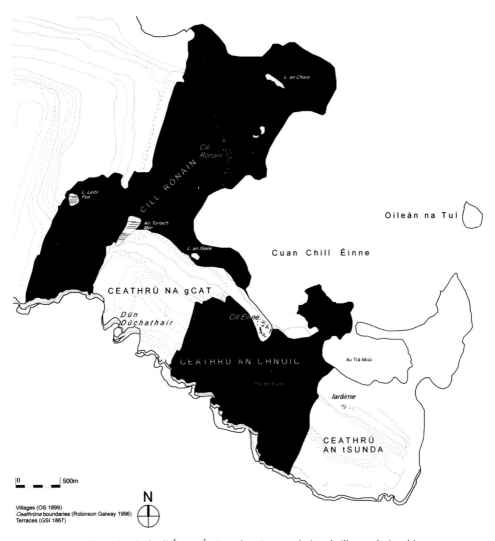

FIGURE 13 Map of townland of Cill Éinne, Árainn, showing *ceathrú* and village relationship.

FIGURE 14 View north over the *ceathrúna* of Eochaill and Mainistir in Árainn.

Cill Éinne is the most easterly townland of Árainn; the name refers to the church of St Enda, the Early Christian saint of Aran. The six *ceathrúna* still exist here: Cill Rónáin, which contains six *leath ceathrúna* (half quarters) farmed by the village of Cill Rónáin; Ceathrú na gCat; Ceathrú an Chnoic (formerly the Hill Farm of Killeany Lodge) and Ceathrú an tSunda (includes Oileán na Tuí). Dún Dúchathair is contained in this townland.

There are two townlands on Inis Meáin, Carrownlisheen and Carrowntemple (in Irish Ceathrú an Leisín and Ceathrú an Teampaill). Given the pattern elsewhere in Connaught, these would seem to be two quarters of one *baile*. However, looking back to the details of the Strafford Survey, it is evident that the *baile* names have changed. Is 'Liske' of the Strafford Survey the antecedent of Carrownlisheen of the Ordnance Survey or contemporary Ceathrú an Leisín? *Les* in Old Irish is the word for an enclosed area in front of a king or lord's house, expressed in modern Irish as *lios*. This word has come to mean 'fairy fort' and is sometimes attributed to ring forts and cashels, apparently because there was a folk belief that fairies dwelt in the former enclosures of the lord's house.[49] Dún Chonchúir (which might have been called a *lios*) is near the modern townland boundary of Ceathrú an Leisín; however, it is not within it. Within the townland and at a small distance from the fort are Loch an Leisín (a small lake) in a place called Gleann an Leisín, Baile an Leisín (a village) and An Leisín (a field).

The puzzle continues if we look at the other *baile* from the Strafford Survey: Kilcannon, which surely must refer to Cill Cheannannach, an early (eighth- or ninth-century) church. But this is not the church of Ceathrú an Teampaill,[50] as it is situated to the far eastern end of Ceathrú an Leisín. It would seem then that Strafford's Kilcannon is today's Ceathrú an Leisin, and Liske is therefore the antecedent of Ceathrú an Teampaill. The *teampaill* must be either Teampall na Seacht Rí, an early church, very little of which remains, or Teampall Mhuire, a fifteenth-century church which was reroofed in 1840 and was in use until 1939, of which nothing remains except the stoup gate and font, which have been moved to the new church opposite.[51] The sites of both of these buildings are in Baile an Teampaill, the village just west of the townland boundary. The neighbouring village just east of the boundary is Baile an Leisín.

The eight *ceathrúna* of Inis Meáin have complicated boundaries that criss-cross each other; the names of all but one of the boundaries are known. Roderic O'Flaherty in 1683 refers to eight quarters in Inis Meáin; however, John O'Donovan, writing about the subject during the first Ordnance Survey in 1839, says that according to the natives of that island there are six quarters and he names them as follows:[52]

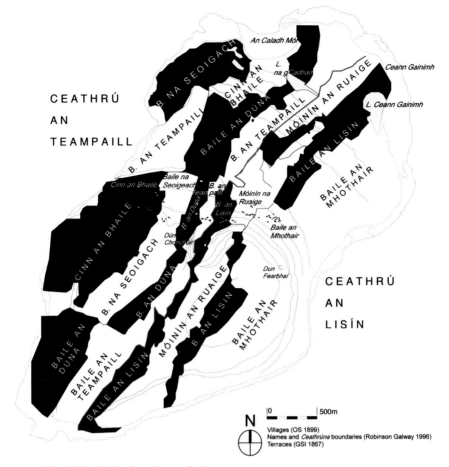

FIGURE 15 Map of Inis Meáin showing *ceathrúna*.

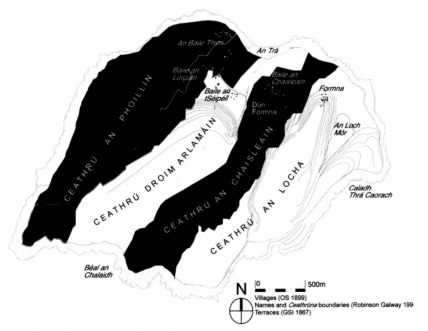

FIGURE 16 Map of Inis Oírr showing *ceathrúna*.

Ceann a Bhaile

Baile an Dún

Baile an Teampaill

Baile an Lisín

Móinín na Ruaige

Mothar

These are the same names today, except that Mothar has now two quarters – Ceathrú na Fearbhaigh and Ceathrú an Iomair – and there is an additional quarter called Baile na Seoigeach. This last *ceathrú* takes its name from a seafarer called Captain Joyce who came to Inis Meáin and settled there.[53] Baile na Seoigeach means 'township of the Joyces'.

The townland of Inis Oírr has the same name as the island. The four *ceathrúna*, boundaries and names are known: Ceathrú an Phoillín, Ceathrú an Droim Arlamáin; Ceathrú an Chaisleáin and Ceathrú an Locha. The most westerly quarter, seems to be further divided into two *leath ceathrúna* (half quarters) on Robinson's map and is farmed by two villages.

Comparing the contemporary townlands and quarters in Aran today with the documentary reference available since *The Compossicion Booke of Conought* of 1585 it appears that, although there may have been changes in names or subdivisions of quarters in some instances, the basic framework remains the same. There is evidence to support the contention that the townland matrix was not largely altered through the centuries. It also raises the question of how it was carried out in the first place. Notwithstanding differences in terminology and size of units between the provinces, the principles of the system exist throughout the country. Yet, the subdivision of the *baile* units, in Aran at any rate, indicate that it must have been divided by some indidivual or group with an intimate knowledge of the land in question and of the manner in which it would be farmed, and with a commitment, for whatever reason, to equivalence between the units. The geographer J. H. Andrews, writing about land measurement in seventeenth-century Ireland, refers to a practice where the youngest heir divided the land and allowed the others to choose their portion, beginning with the eldest.[54] While this practice refers to the division required in the inheritance of a farm, it may be that at an earlier date a similar method was used to subdivide the *baile* unit. Sir Henry Piers, writing in 1682 about Gaelic farming in its final stages in Westmeath, suggests that the ability to judge the value of land in terms of its fertility seems to have been commonly held in Ireland. He gives an elaborate description of how land is divided and lots are cast for each share, and how men will take upon themselves 'to be judges to an extreme nicety of the quality and quantity of each rood of ground; and, to make sure work, will bring their ropes to measure, as formally as a surveyor his chains.'[55]

In Aran the land-division units of the townlands and the *ceathrúna* that are bounded by drystone walls often follow the fissures of the limestone pavement, particularly on Árainn, the largest island, where the divisions run mainly north–south, though there are some interesting exceptions. The typical boundary wall of the *ceathrú* starts high at the Atlantic cliffs and traverses the island in a northward direction following the line of the joints along the limestone pavement, sometimes turning to follow an inland cliff, stepping down across the north-facing terraces as they reach the shallower northern coast. The land units have a long and narrow form to take in different types of land and to give each *ceathrú* access to the shore. Each *ceathrú* contains: land on Na Craga ('the crags'), the high limestone plateau towards the Atlantic cliffs where cattle are wintered; land on the sheltered side of the lower fertile shale terraces and limestone pavements, which are endowed with wells and springs; land at the north coast where the terrace steps down to sea level and sometimes contains part of a sandy beach.

Sand was an important ingredient when 'making land' on the limestone pavements, which occurred throughout the nineteenth and twentieth centuries. Sand was drawn for this purpose from the beaches at Cill Mhuirbhigh and An Trá Mhór in Árainn, Ceann Gainimh in Inis Meáin and the beach along the northern coast of Inis Oírr. Most *ceathrúna* contained or had a share in a natural harbour where seaweed could be harvested. Seaweed was the other crucial ingredient for the reclamation of land. The *ceathrú* also contains or gives access to a freshwater lake, as at Loch Dearg in Eoghanacht or Loch Phort Chorrúch, shared between Ceathrú Chorrúch and Ceathrú Bhaile na Creige on Árainn, and An Loch Mór on Inis Oírr.[56] Another source of fresh water is the *tuar loch* (turlough), a lake which emerges and disappears with fluctuations in the water table.[57]

An interesting exception to the north–south rule is at the boundary between Ceathrú Chill Mhuirbhigh and Gort na gCapall. Here, the boundary divides the fertile land on the sandy soil of the northern coast between the two *ceathrúna*, giving each a fair portion. It is clearly visible today, separating the small enclosures of grassy fields known as An Caiseal, in Gort na gCapall, from the large fields of the Kilmurvey House farm, cleared of stones and walls. This *ceathrú* wall does not appear to follow any natural boundary. An extra triangle of land bounding the northern coast and the sandy beach of Cill Mhuirbhigh is also part of Gort na gCapall. Without this piece, which provides access to seaweed and sand, the *ceathrú* would have been unsustainable. This particular boundary between the small farmers of Gort na gCapall and James O'Flaherty of Kilmurvey House, son of the largest tenant and middleman Patrick O'Flaherty, became the

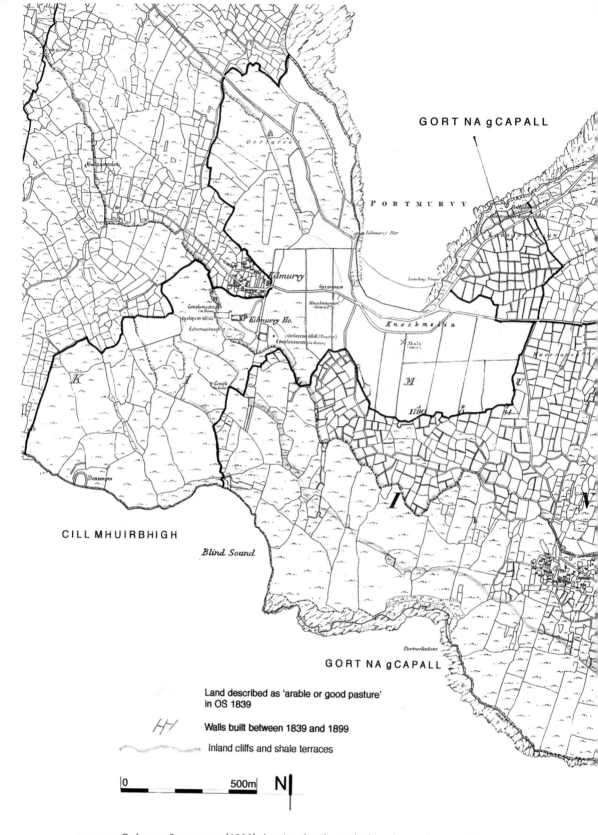

GORT NA gCAPALL

PORTMURVY

CILL MHUIRBHIGH

Blind Sound

GORT NA gCAPALL

Land described as 'arable or good pasture' in OS 1839

Walls built between 1839 and 1899

Inland cliffs and shale terraces

0   500m   N

FIGURE 17 Ordnance Survey map (1899) showing details overlaid on the *ceathrúna* of Cill Mhuirbhigh, Gort na gCapall and Corrúch in Árainn. Reproduced by permission of the Board of Trinity College Dublin. Sheet No. 110.

CORRÚCH

interface of the Land Wars of the late nineteenth century. In a discussion about equivalence between units Gort na gCapall would also feature, as would the narrow, rocky Baile na Creige. The former is much larger than the other units (more than twice the size of Cill Mhuirbhigh or Fearann an Choirce) and also has more fertile land than the others. Perhaps the two lost *ceathrúna* of the Cill Mhuirbhigh townland are contained somewhere within it? If so, it was not revealed to Tim Robinson as such in his perambulations with local people to ascertain the boundaries and names for his map of the islands. Robinson locates the lost *ceathrúna* somewhere in Ceathrú Chill Mhuirbhigh and Ceathrú na Carraige.[58] The latter, Baile na Creige, is the narrowest *ceathrú* on Aran and the one with the least fertile land. It is also linked only by a road to that portion to the north which gives it access to the fertile lower terraces and the seaweed of the northern coast; however, the quarter includes almost the whole of Loch Phort Chorrúch. Whether it was laid out like this with less of everything compared to the others, or parts of it were nibbled away over time, cannot be said. Not surprisingly this *ceathrú* was associated with poverty, or at least a place called Baile na mBocht or Baile na Sean – the village of the poor or the village of the old.

On Inis Meáin, the townland boundary clearly divides the island in two, not always following the line of the sets of joints. It cuts through the centre of the ravine in the middle of the island, giving an equal share of terrace and shale to both townlands. It cuts across the crags on the high ground and the lower bare limestone pavement of the northern part of the island on an equal basis, and veers to the northeast at the northern coast to divide equally the precious sandy terrain, which has supported fertile fields at least since the early nineteenth century and probably before. The *ceathrúna* boundaries are complicated and result in dispersed parcels in an apparent effort to give each *ceathrú* the necessary types of land for a sustainable unit. They are often parallel to the townland boundary and cut across the inland cliffs, parcelling out the fertile shale terraces among the *ceathrúna*. The field-boundary walls are generally perpendicular to the line of the inland cliffs, establishing a pattern of small fertile fields at the base of the cliff, gradually spreading out over the terraces as reclamation of the limestone pavement occurs.

On Inis Oírr, the interaction between the land-division system and geology or terrain is also very clear. The central boundary between Ceathrú Dhroim Arlamáin and Ceathrú an Chaisleáin divides an Creig Mhór at the highest point of the island. It then cuts down northwards through the ravine below Dún Formna, known as An Ghleann, to divide the shale terraces and, veering to the northeast at the sandy beach, equally divides the resources of sand and seaweed at the northern coast. The boundary across the crag does not parallel the joint system as in Árainn, but is parallel to the inland terraces that face northwest and provide the fertile fields of Ceathrú an Phoillín, the most westerly quarter. The other *ceathrúna* boundaries are roughly parallel. The island is thus divided into four, ensuring that each *ceathrú* has a fair share of shale terrace, crag, sandy beach and access to the sea. Ceathrú an Loch Mór has, in addition, the wonderful triangular-shaped freshwater lake carved out of the ravine by rainwater erosion in geological time.

The *ceathrú* is divided again into cartrons and then, on Aran, into *cnagaire*, anglicised as croggery. *Cnagaire* is a word connected with a fourth measure in Irish. The old Aran Irish for the word is *crangaire*. (There is a place in Fearann an Choirce on Árainn called An Crangaire.) Pádraig Ua Duinnín's dictionary, *Foclóir Gaedhilge agus Béarla*, first published in 1904, translates *cnagaire as*: '16 English acres (Aran)', while that of Niall O Dónaill (1977) gives: 'a measure of land, equated to 16 acres, a small holding'. This word may be unique to Aran.[59] During the law courts in Cill Rónáin, Árainn, at the end of the nineteenth century to assess the question of high rents, Oliver J. Burke, a barrister for the Land Commission, who was present, gives this account:

> Mr. Thompson, of Clonskea Castle, county Dublin, sworn. Is the agent on the estate; succeeded his father, who had been agent for many years: 'The lands are in the hands of the tenants, with the exception of two croggeries which are in my occupation.'
>
> The learned chairman, Mr Crean., BL, inquired what a croggery meant. Witness said that a 'croggery' was a very ancient name for fourths. The entire islands were divided into townlands, which townlands contained 4 or 6 quarters each, every quarter containing 16 croggeries, and every croggery containing 16 acres. Inishmore thus contained 4 townlands and 4(t) x 6(qrs) x 16(crog) x 16(acr) = 6,144 acres. On Inishmaan there are two townlands which contain 6 quarters each. On Inisheer there is only one townland containing 4 quarters. The tenants have manure and seaweed from the seashore free of charge.[60]

This suggests that the *cnagaire* (croggery) was peculiar to Aran, if the 'learned chairman' – who one assumes was dealing with land matters in other locations – did not understand the word. Even today some islanders use the term *cnagaire*, cartron and half-cartron rather than acres to describe their landholding.[61] Within the *cnagaire* the land is divided again into fields and smaller fields.

All of the boundaries of these divisions and subdivisions – from the townland boundary to the smallest field – are made with drystone walls, the material gathered from the glacial debris scattered around the surface of the landscape and constructed by the islands' inhabitants, stone by stone. Because of the density

FIGURE 18 Ordnance Survey map of Inis Meáin, surveyed 1899, terraces from the GSI map added. Reproduced by permission of the Board of Trinity College. Sheet No. 119.

L

FIGURE 19 Ordnance Survey map of Inis Oírr, surveyed 1899, terraces from the GSI map added, *ceathrúna* boundaries added. Reproduced by permission of the Board of Trinity College. Sheet No.s 119, 120.

of the stone-wall boundaries and the extent of reclaimed land it comes close to being a built landscape. It is a landscape largely constructed by human beings in the context of an exceptional natural landscape, which records and displays the geology, the history and the societal patterns of its 4,000-year-long human habitation. This densely patterned landscape mosaic, rendering the fissures of the geological joint pattern manifest and recording the history of the islands in a less immediately apparent way, has become the characteristic image of the cultural landscape of Aran.

CHAPTER 3

Settlement and Landholding

Village and *Ceathrú*

Clachan and Rundale?

Fertility and Land Reclamation

FIGURE 20 Village of Eochaill in Árainn.

Village and Ceathrú

A number of patterns of settlement can be discerned in Aran reflecting the layers of inhabitation of the islands through the centuries: from the early 'forts' or cashels and the fragmented habitations of early Christian monastic communities to groups of clochans of unknown date. There is also dispersed settlement such as the medieval tower house of the O'Briens, built within the walls of Dún Formna on Inis Oírr, or the two larger farmhouses on Árainn: Killeany Lodge and Kilmurvey House. However, the dominant layer of settlement and that which is most relevant to the field-boundary system that is the subject of this book, is the layer containing the villages of Aran.

In all, there are fourteen villages on Árainn, eight on Inis Meáin and five on Inis Oírr. Each village farms the land of one or sometimes two *ceathrúna*. The labour-intensive work of building the drystone field-boundary walls and the reclamation of land from the rock was carried out by the inhabitants of the villages that were attached to the *ceathrúna*; the field systems and settlement pattern together form a human habitat.

Various interdependent factors have contributed to the location and morphology of these villages, including the geology of the islands, the self-contained economic land units of the townland matrix – the *ceathrúna* – and earlier settlement nuclei, but possibly the most fundamental influence is the pattern of hydrology created by the primary geological formations described in Chapter 2.

The hydrology of Aran is determined by a combination of the almost horizontal layering of limestone and shale beds and the vertical fissures in the limestone pavements that penetrate the rock bed. Rainwater filters down through these joints and builds up in the limestone bed because of the impermeability of the shale layer at the base. As the water table rises, numerous springs emerge along the face of the high Atlantic cliffs (where it is of no use to humans) and also more sparsely at the bottom of the inland cliffs. It is here, in the shelter of the terraces, below the escarpment where there is protection from the Atlantic gales and the prevailing southwest winds, and in close proximity to the collection points of fresh water that has filtered through the limestone bed, that human settlement occurs.

This pattern of settlement is related to the location of freshwater sources. In general, the village is located near a grouping of springs, usually above the line of the emergence of water, but sheltered from the prevailing winds by a higher terrace. During periods of heavy rainfall in winter, these areas that are dotted with springs are sometimes flooded. A good example of this is at Gort na gCapall in Árainn. The village has been built on the cliff above An Sruthlán, a fertile glen full of springs, which becomes a lake after heavy rains.

The shale layer, as well as being the point of emergence of fresh water, also supports fertile ground. These sheltered areas, protected from the erosion of soil that is common on the limestone pavements, are where the small gardens and field enclosures surrounding the villages begin. It appears that over the centuries the process of enclosure moved out from the base of the inland cliffs to enclose the manmade fields of the limestone pavement.

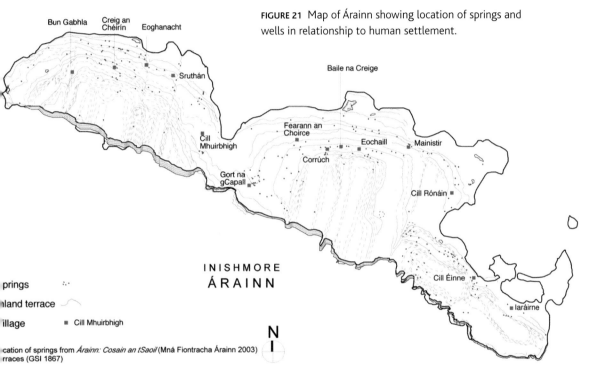

FIGURE 21 Map of Árainn showing location of springs and wells in relationship to human settlement.

Bun Gabhla   Creig an Chéirín   Eoghanacht

Sruthán

Baile na Creige

Fearann an Choirce

Cill Mhuirbhigh

Eochaill   Mainistir

Corrúch

Gort na gCapall

Cill Rónáin

INISHMORE
ÁRAINN

Cill Éinne

prings

land terrace

illage   ■ Cill Mhuirbhigh

Iaráirne

cation of springs from *Árainn: Cosain an tSaoil* (Mná Fiontracha Árainn 2003)
rraces (GSI 1867)

N

FIGURE 22
An Sruthlán in January, after heavy rain. Some of the houses of Gort na gCapall can be seen to the right of picture.

On Aráinn and Inis Meáin, the villages generally occur at a central point of the *ceathrú*, close to a sheltered and fertile terrace, below the limestone pavements, Na Craga ('the crags'), where wintering of cattle occurs, and above the summer pasture of the lower terraces.

Bóithre (roads), which are generally small green roads enclosed by drystone walls, start at the village and stretch out northwards and southwards, giving access to the dispersed patchwork of holdings across the *ceathrú*. They also lead to the coast, from which seaweed could be harvested, sand could be drawn, and boats launched for the supplementary activity of fishing. The Aran farmer was a *fear talamh agus trá* ('a man of land and strand').

On Inis Oírr, the positioning of villages is slightly different because of the topography of the island. The high ground of the upper terraces comes up close to the northern coast; the shale terraces then step steeply down to the sandy beach, An Trá. The jointing system, as described earlier, has cut into the limestone beds, forming deep ravines, leaving three 'headlands'. The villages are situated in relation to these, enjoying the shelter and the fertile shale terraces. The most easterly and highest village, perched above Loch Mór, is called Formna (which means 'terrace') and the lowest and most westerly is called An Baile Thíos or An Baile Thiar ('the town or village below', or 'the village to the west').

In the east–west direction, settlement has occurred in direct relationship to the *ceathrúna*. In some instances, *ceathrúna* boundaries may have been created in relation to an already existing settlement. (Maybe this could explain why on Árainn there are six *ceathrúna* to each townland, when one would expect there to be four.) When the village farms two *ceathrúna* its location often straddles the boundary, as in Creig an Chéirín in the townland of Eoghanacht.

It is worth noting that each of the seven major 'forts' or cashels of the three islands is located in a separate *baile* or modern townland. It is possible that, as there is today a relationship between village and *ceathrú*, there was also once a relationship between cashel and *baile*. The archaeologist Etienne Rynne considers this possibility in an essay about the forts: 'Each of them is to be found within a different townland, townlands which are topographically distinct and which may owe their origins to earlier Celtic tuatha; each Aran "fort" should be looked upon as the temple of the local tuath, just as each parish nowadays has its own church and each political entity (town, city, county, state) its own meeting-place (town-hall, mansion house, county buildings, parliament).'[1]

In the largest island, a study of the relationship between village and *ceathrú* reveals that some of the villages occur close to earlier settlement sites. In the townland of Eoghanacht, the village, also called Eoghanacht, occurs beside the early monastic settlement of Na Seacht dTeampaill ('the seven churches') which included domestic buildings. The village straddles the boundary between Ceathrú an Turlaigh and Ceathrú an Ocht. The villages of Cill Mhuirbhigh, Corrúch, Mainistir, Cill Rónáin and Cill Éinne, all occur close to early Christian churches and take, in some cases, their names from them.

Clachan and Rundale?

We refer to the settlement form as 'village', but it is not of the English or Norman village type, which is centred around a green or church, can be formally planned, and belongs to the medieval feudal tradition. The islanders use the word *baile* when talking about their village in the Irish language (not to be confused with the earlier use of the same word meaning 'township'). In terms of morphology the villages have the characteristics of what the influential Welsh geographer Estyn Evans, working in the mid-twentieth century, and the school of Ulster geographers who followed him would have called 'clachans'. An inspection of the first-edition Ordnance Survey map of Aran, surveyed in 1839, shows that the land surrounding most of the villages is fertile and enclosed in small fields. The outlying parts of the *ceathrúna* are less enclosed, particularly on the land of the crags up towards the southwestern cliffs. This land, used for wintering cattle, may have been held in common, and gradually became enclosed and allocated to individual holdings albeit dispersed throughout the *ceathrú*. This pattern is similar to the rundale and clachan system described by Evans and other Ulster geographers. While at first glance, in the nineteenth-century Ordnance Survey map, the villages appear to match the description of typical clachans: houses sometimes parallel and in an informal layout without a church or focus building of any kind. However, a closer inspection reveals a more complex picture.

The word clachan literally means 'a thing of stone', and its use has come under criticism in recent years, although it does seem apt in Aran. Evans, and others such as Desmond McCourt, saw in clachan settlements of the nineteenth century a pattern of community farming and settlement that had roots in the medieval and even the Neolithic period. Their field research was primarily along the Atlantic fringe in northwest Donegal and north Connaught, and they concluded that this type of peasant settlement had occurred across the country before being disrupted by a succession of invaders. In their view, the clachans were ethnically determined, and they supposed the inhabitants to be descended from the pre-Celtic people.[2] This view has been challenged in more recent times, and the historian and geographer Kevin Whelan in particular has pointed out that nineteenth-century clachan settlement in Connemara was new rather than old. The clachans developed under population pressure in the pre-Famine decades, particularly in areas where rough or waste land was available for cultivation and where there was ready access to manure in the form of seaweed, such as along the Connemara coast. Whelan forcefully declares that:

> They are not the degraded relics of an archaic, aboriginal settlement form, practising primitive agriculture in 'refuge' areas. They are instead a sophisticated solution to specific ecological, environmental and social problems, which maximised the carrying capacity of a meagre environment in an expanding demographic regime.[3]

There are parallels in Aran where there was plenty of rocky land that could be reclaimed through hard work. There was ready access to seaweed and sand, an understanding of the reclamation method, and a growing population that would provide the labour force. However, Aran was not an area of new settlement; as described earlier, it had been farmed and inhabited from prehistoric times, and the traces of earlier settlement can be seen in today's villages.

A village such as Gort na gCapall seems to have evolved from a single farmstead following a pattern seen elsewhere by the geographers of the Ulster school. McCourt stressed the dynamic nature of Irish rural settlement and holds the view that clachans could develop from a single farmstead and return to one again, depending on economic and social conditions. McErlean, writing about the land-division system, also points out that the matrix did not imply a static society and that settlement could wax and wane within the boundaries of the land-division units.[4] It is worth remembering that although the villages are not formally planned in relation to geometry or to a focal point, they are very precisely planned in relation to topography, sources of fresh water and fertile land. The writer Tom O'Flaherty, brother of the more famous short-story writer Liam O'Flaherty from Gort na gCapall, says that his village was founded by an ancestor of his own, Bartholomew.[5] Apparently, their father used to recite his genealogy as follows, when asked to trace his descent from this founder ancestor: Maidhc Mhicil Phádraic Bheartlaiméid Bhriain Bheartlaiméid.[6] By rough calculation, this Beartlaiméid Ó Flaithbheartaigh was born sometime in the seventeenth century and perhaps founded the village towards the end of that century, contemporaneous with Roderic O'Flaherty's account of the islands.

The village's natural harbour – the only one on the high Atlantic cliff side of the island – is called Port Bhéal an Dúin, ('the Harbour at the Mouth of the Fort'), the fort being Dún Aonghasa. Standing in Gort na gCapall today, observing fishing nets and lobster pots thrown over the walls outside the houses, the looming fort on the cliff beyond and the constant sound of the sea, one senses that the place has retained its aura of an epic landscape, and one is inevitably led into speculation. Perhaps the extraordinary *chevaux de frise* that surround the outer rampart of Dún Aonghasa, rather than being a protection from armies on horseback, is in fact protecting the 'sacred space' from the wandering horses of Gort na gCapall ('field of the horses') and indeed the cattle and horses of the rest of the island.

A survey of villages carried out in 2003 showed that of the thirty-eight houses in Gort na gCapall fourteen are owned by people with the surname Ó Flaithbheartaigh (O'Flaherty)[7], descended perhaps from the original Beartlaiméid Ó Flaithbheartaigh?

Under the Gaelic system farming was carried out in common (not collectively), with a complex legal apparatus governing the maintenance of roads, boundaries, sharing of commonage areas etc.[8] Various exactions were due from each *ceathrú* to the Gaelic lord. It is possible that each *ceathrú* may have been cultivated in common by a kin group in the medieval period or even more recently. The adjustments necessary to accommodate the new system of post-plantation Ireland probably occurred slowly over time. Research elsewhere has shown that in times of difficulty and population pressure, people reverted to the communal method of farming.

The transition to freehold and tenant status was not the end of communal farming in Ireland. It could still be found in several areas at the time of Arthur Young's visit in the late eighteenth century, and traces of its survival have been studied in detail for County Tipperary by Ingeborg Leistser (1976). The communal farming tradition took on a new life with the spread of clachans across the west and northwest in the eighteenth and early nineteenth centuries.[9]

Even today, the pattern of dispersed holdings within the *ceathrú* has traits associated with communal farming, where each farmer shares boundaries with every other member of the village. In the twentieth century some agricultural work in common was done. For example, in the village of Eochaill each family would help to gather and harvest the kelp, which was then bartered for a supply of turf from Connemara. This would provide the village with fuel for the winter;[10] the same was probably true of other villages. The islands lacked their own source of fuel, having neither bog nor trees. This dependency on Connemara for an essential resource initiated a considerable commerce based on the turf trade.[11] Potatoes, herrings and other fish, sally rods and feathers are some of the goods that were exchanged for turf which was conveyed to Aran by hooker, the traditional sailing boat of the west coast.

By whatever name the villages are called they seemed to exhibit aspects of the clachan and rundale system; in particular the apparent formlessness of the layout of the village and the dispersed holdings within the land unit, which is a vestige of communal farming. They are clearly intimately linked to the land-division system and to earlier settlement nuclei which suggests that they have roots much further back into the past than the nineteenth-century clachans of Connemara and other areas of wasteland, where groups of people settled in times of hardship and population rise. What they share with these is the method of reclamation of land and the layout and construction method of the houses. However, to paraphrase Whelan but to draw a different conclusion in the case of the villages of Aran, it seems that they are both relics of an ancient settlement form and a sophisticated response to various environmental challenges in a time of demographic expansion.

Fertility and Land Reclamation

From this settlement form, laid out informally but in response to climate (shelter) and topography (sources of water and fertile land) and located centrally in the *ceathrú*, the inhabitants could farm their land unit and build walls: to establish boundaries; to clear the land; to promote intensive agriculture on scarce grassland; and to protect livestock and crops from Atlantic gales. However, the matrix also allows for other types of settlement and it is worthwhile in the search for answers about intensive wall building to look at the exceptions to the rule.

On the islands there are two instances where the settlement pattern in recent centuries took a different form. Compare the *ceathrú* of the Hill Farm of Killeany Lodge (Ceathrú an Chnoic) or the Kilmurvey House farm (Ceathrú Chill Mhuirbhigh) to the rest of the landscape. These are single houses dispersed in the landscape, that is to say, they are not in the village, and they are larger than the village houses. In both cases a single family farmed the land of one *ceathrú* or, at certain times in the past, more than one. The landscape appears in stark contrast to that of the neighbouring *ceathrúna* of Árainn and the two smaller islands because of the absence of subdivision. Their fields, in comparison to the multiplicity of small fields of their neighbours, are large and empty of stone walls.

Of these two houses, the former is an early nineteenth-century one-storey farmhouse with outhouses and walled garden, situated on high land above the village of Cill Éinne. The present house may have been built on or contain the remains of an earlier house and home of the Fitzpatricks who farmed the islands in the early eighteenth century. By 1820 the tenant of Ceathrú an Chnoic was a family called O'Malley;[12] later in the century the land was leased from the absentee landlords (the Digby family) by the O'Flahertys of Kilmurvey House. In the Ordnance Survey maps of 1839 and 1899, the *ceathrú* has an open unenclosed aspect. By 1927 it had been divided by the Land Commission among people from the neighbouring villages of Iaráirne and Cill Éinne, which required the building of boundary walls. Consequently, today this *ceathrú* has more enclosure than shown on the 1899 Ordnance Survey map, including some drystone walls, which are very recently built. However, it lacks the densely webbed and enclosed landscape that is characteristic of the rest of the islands, because by the time the ownership had been transferred to a number of owners, the necessity or customary practice of intense subdivision of land was over.

The latter is a nineteenth-century two-storey house (incorporating an older one-storey house), a short distance from the village of Cill Mhuirbhigh, built by James O'Flaherty, son of Patrick O'Flaherty. O'Flaherty first leased a *ceathrú* and a half at Kilmurvey from the Digbys in 1812,[13] but the family already had established themselves in Kilmurvey at that stage. O'Flaherty was a 'middleman': that is, he sublet land to other tenants as well as farming some of the land that he leased himself. He became the largest leaseholder of land on Aran during the nineteenth century. Tracing Ceathrú Chill Mhuirbhigh in the Ordnance Survey map of 1839 and 1899, it is clear that, in contrast to the pattern elsewhere, the fields have become larger and some field boundaries have been removed. Throughout the nineteenth and twentieth centuries this *ceathrú* was farmed by a single family; even today it is a *ceathrú* of large fields and few boundaries and continues to be farmed by the family that lives in Kilmurvey House. Although the days of the middleman in opposition to the smaller landholders are over, the cultural landscape of the *ceathrú* continues to manifest a different history.

Elsewhere in the Aran landscape the land is minutely divided. Many of these divisions are boundary walls. Since each landholding is dispersed throughout the *ceathrú* there are more boundary walls than would be necessary if each holding were consolidated. Yet, even within the holding, the land is divided. Walls were built to clear the land of stones, to provide shelter, and to create enclosures so that each field could be intensively grazed before moving animals to the next field. However, the two large leaseholders of Kilmurvey House and the Hill Farm obviously found no need to use this method of intensive farming, nor apparently did they require this kind of shelter for their animals. The Kilmurvey farm is in a sheltered area and the land is described as fertile in the 1839 Ordnance Survey map. Hence walls for shelter and reclamation may not have been necessary. This is not true of the Hill Farm, large sections of which are limestone pavement that has neither been reclaimed nor cleared of stone, but the *ceathrú* also enjoys some shelter and fertile land along the northern coast. The comparison suggests the obvious conclusion that a *ceathrú* farmed by many people results in many stone walls. In addition, it presumes that the larger landholder in the case of Cill Mhuirbhigh had the resources to clear the land of stones and remove them from the *ceathrú*, and that stone walls as a buffer against the wind were not required, because it was in a sheltered area.

If wall building occurred as a result of demographic pressure on the land then it is likely to have happened in the nineteenth century when the population rise in the rest of Ireland was reflected in the population of Aran. Medieval historians believe that the population of Ireland in the twelfth century may have been as low as 500,000. On the eve of the Great Famine of 1845 it was 8 million, rapid population growth having occurred at the end of the eighteenth and the beginning of the nineteenth century.

FIGURE 23 Ordnance Survey map of 1839 (facing page), Ordnance Survey map of 1899 (this page), showing changes in field boundaries around the villages of Corrúch, Baile na Creige and Eochaill in Árainn. Reproduced by permission of the Board of Trinity College Dublin. Sheet No. 110.

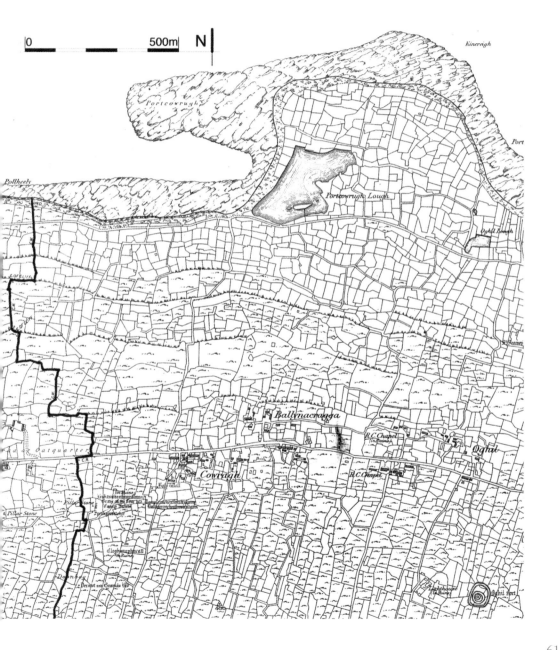

The earliest cartographic reference available showing townland, *ceathrúna* (the boundaries are shown though not named) and field boundaries is the first Ordnance Survey map (1839). The maps of the Strafford Survey (1636), lost since 1711, may have shown townland and *ceathrú* boundaries. An examination of the Ordnance Survey maps of 1839 and 1899 reveal that an intensity of wall building occurred in the second half of the nineteenth century. This assumption is based on the reliability of the maps in relation to field boundaries and a word must be said about this. During the first years of the Ordnance Survey not all of the maps included fences. In the North, six counties had been mapped without field boundaries before Lt. Colonel Thomas Colby, director of the Ordnance Survey, gave an instruction in November of 1835 that 'leading fences should henceforth appear on the plans'.[14] However, it took some years before this order was followed to the letter, but by 1842 when the survey had reached the southern counties every fence was measured, drawn and engraved. Aran was mapped in 1839. A list was published in 1849 which gave the areas to be revised in order to bring them in line with the southern counties. Galway was considered to be nearly as complete as the southern counties.[15] In the 1839 map there are many field boundaries shown throughout the three islands, in 1899 many more were added. The question is whether we can accept that the 1839 map shows the true picture of the field-boundary system at that date. Of course, if field boundaries appearing in 1899 were actually there in 1839, they are all the more remarkable for that.

While we cannot be sure that a boundary shown on the 1899 map was physically there in 1839 but not drawn, some assurance is felt by comparing an instance where the opposite to the general rule occurred: field boundaries were removed in the 1899 map. There is, for instance, a change in the map which is reflected in a story about an event of the time. In the west of Árainn in the townland of Eoghanacht there is an example where more field boundaries occurred in the earlier map than in the later one. The later map shows the southern half of Ceathrú an Turlaigh as a dense web of enclosed fields, but the small enclosures surrounding the cashel itself and those to the north of the cashel which appear on the earlier map are gone. These have been replaced with large rectilinear fields, similar to those of the O'Flaherty farm at Ceathrú Chill Mhuirbhigh. This change in the map is reflected in a story recounted by Robinson which refers to an event in the mid-nineteenth century, when the small landholders of Ceathrú an Turlaigh, the *ceathrú* immediately to the west of Dún Eoghanacht, were evicted from their lands in order to make way for a lease

Facing page: **FIGURE 24** Ordnance Survey map of 1839 (left), Ordnance Survey map of 1899 (right), showing changes in field boundaries around Dún Eoghanachta in Árainn. Reproduced by permission of the Board of Trinity College Dublin. Sheet No. 110.

0   500m   N

to Patrick O'Flaherty of Kilmurvey.[16] Tom O'Flaherty also mentions this event when writing in the early twentieth century[17] about the 'good land around Dún Eoghanacht' having been taken by the 'land grabber' who had people evicted from their lands so that he could have it for himself. The map therefore reflected an event of which we have a verbal record, the existence of small fields in 1839 and their removal by 1899.

If we take the date of the second-edition Ordnance Survey in 1899 and look back to 1839, the most striking thing about the comparison between the two maps is the intensity of wall building during the period, the amount of land reclaimed, particularly on the north-facing terraces, and the fact that, as well as reclamation, the fertile land is almost everywhere subdivided into four or six smaller fields. Walls were built, too, on the limestone crags of the southwestern flank but with lesser intensity. At the coast reclamation was easier than elsewhere; seaweed and sand – the ingredients of reclamation – were within reach. However, the practice of 'making land' on the limestone pavements high above the source of the fertilising materials is likely to have occurred only as a response to extreme population pressure and harsh economic conditions.

In the sixty years between the maps the population increased marginally and the number of houses in the villages, in general, did not increase, yet this is a period of intense activity. We know from historical accounts that the Great Famine of 1845–1849 touched Aran lightly, but the second half of the nineteenth century was a time of difficulty and food shortages.[18] Robinson suggests that reclamation only became important when the indigenous soils under the scarps and of Na Craga were unable to support the food demand.[19] Clearly, the new fields of the terraces were planted with potatoes to provide a quick food supply and also because the sowing of root crops or potatoes was part of the land-making process and helped the soil to get a foothold.[20] A story told to Robinson by an islander shows that fields near the sea were cultivated for potatoes around the time of the Great Famine, because the blight affected them less there.

> And isn't it an odd thing that the gardens nearest the sea were the soundest. Dónall Mhicil noticed this, and that year he sowed potatoes in bits of land he had all the way from Corrúch shore west to An Duirling Bhán. Patches here and there, only four or five ridges in each, perhaps, but all the same he had the most potatoes in this village or in the island if it comes to that. There wasn't so much as one rotten potato among them. Fine healthy potatoes.[21]

Facing page: **FIGURE 25** Ordnance Survey map of 1839 (above). Ordnance Survey map of 1899 (below), showing changes in field boundaries around the village of Iaráirne in Árainn. Reproduced by permission of the Board of Trinity College Dublin. Sheet No. 119.

FIGURE 26 Ordnance Survey map of 1839 (this page). Ordnance Survey map of 1899 (facing page), showing changes in field boundaries around the village of Cill Rónáin, Árainn. Reproduced by permission of the Board of Trinity College Dublin. Sheet No. 111.

Statistics of the time show that most of the land under crops was used for potatoes during these years, increasing from an overall figure of 792 acres in 1865 to 1,200 acres in 1898.[22] This period seems to have been more difficult than the first half of the century, and is reflected in the activity in the landscape.

In the 1839 Ordnance Survey map, there is a clear pattern of arable and good pastureland at the base of the inland cliffs across the three islands. In the lee of the cliff, the shelter that attracted settlement also helped to protect soil and shale from erosion. In addition, these areas with their abundance of wells and freshwater pools were important for stock rearing and crop growing. As one might expect, this is also where reclamation began; indigenous fertile land close to the cliff was extended northward through reclamation until the whole terrace was a fertile plain of enclosed fields. This pattern of reclamation is evident from the 1899 Ordnance Survey map.

The first-edition maps describe land in two categories: good pasture/arable land and rocky bad land. The surveyors' notebooks describe land in three categories: bare rock, rocky pasture and arable. A close study of the survey books may reveal a clearer picture of soil fertility. Quite a small amount of land is described as 'good pasture or arable', less than one-third for Árainn. In Inis Meáin, for example, approximately 85 per cent is described as rocky pasture. It is possible that the nineteenth-century surveyors were not aware of the value of rocky pasture on the crags in Aran. They may have considered this 'rocky bad land', but in fact it was valuable land for wintering cattle. Young cattle from Aran and the Burren were sought by graziers throughout the nineteenth century.[23]

A description by J. T. O'Flaherty, who visited the islands in 1824, gives an account of the agriculture. He describes crops such as potatoes, rye and black oats being grown, as well as small quantities of barley, wheat and flax. He describes the pastures stocked with sheep, goats, small cows and horses and: 'The mutton is considered to be delicious; but their most profitable stock consists of calves, which are reputed to be the best in Ireland.'[24]

Agriculture is supplemented by fishing and kelp-burning. There are 120 boats between the three islands which take in 'immense quantities of cod, ling, haddock, turbot, gurnet, mackerel, bream, etc, and in the season, abundance of lobsters, crabs, scollops, cockles, mussels, etc'. He also describes the season of fishing for herring and sunfish, and 'within forty miles of the coast is the great cod-bank, which is supposed to reach Newfoundland'.[25] The gardens, he says, 'are well supplied with every necessary vegetable, and the isles abound with a variety of medicinal and sweet herbs'.[26] O'Flaherty also remarks that the population is dense, (more dense than the rest of Ireland?)

J. T. O'Flaherty's account is roughly contemporaneous with the first Ordnance Survey, and both map and documentary evidence seem to describe a modest but sufficient way of life and an agriculture that was varied and potentially viable. It seems to be in contrast to what was happening in Ireland as a whole, particularly along the west coast, where the entrenched political and economic conditions that precipitated the ubiquitous growth of the potato had eroded all other forms of agriculture among smallholders. The strength of Aran was a reasonably diversified agriculture, in addition to fishing and kelp-burning, to sustain a meagre economy. Nonetheless, there were minor famines on Árainn in 1822 and 1823, followed by cholera epidemics which caused loss of life.[27]

Less than ten years before the Great Famine, the islands appear to have had communities living in farm villages surrounded by enclosed fields, which they farmed individually. The crags may have still been farmed in common for wintering cattle, and unreclaimed rocky pastures on the terraces may also have been held in common. The population was rising steadily as elsewhere in Ireland: 2,732 on the three islands in 1821, and considered dense by J. T. O'Flaherty in 1824. It is not possible to say definitively when the land was reclaimed or the walls of the 1839 map were built; nonetheless, it is likely that as the population rose, and both the necessity and the workforce required for intensive labour existed, reclamation and enclosure occurred. These new fields were then added to individual holdings and the amount of commonage was reduced.

This process is still happening today: a farmer registers a piece of land in his own name that was part of the earlier commonage of the village, without opposition from the neighbours. Legally the other inhabitants of the village have a share. However, they accept that the land belongs to him because he and his family reclaimed it and built the walls, in agreement with their neighbours.

While we have established that the framework of fractional land-units, marked by the *ceathrúna* boundary walls, are likely to have existed since the twelfth century and the *baile* boundaries (the modern townland on Aran) are even earlier, we have no evidence that the physical boundaries – those seemingly flimsy and precarious drystone walls – that mark the divisions today, date from that time. We do know that elsewhere in Ireland, such as the Céide Fields in present-day County Mayo, field systems enclosed by stone walls or stone ditches not unlike the landscape of Aran, occurred as early as the Neolithic period. Studies of stone field-boundary walls in the similar landscape of the Burren have shown that some 'mound walls' were constructed at least 2,000 years ago.[28] However, the boundary walls of Aran await archaeological investigation. Drystone walls cannot be carbon dated because of the lack of mortar, but lichen supported by the walls could be used to date them.[29]

With respect to the fertility of the land and the necessity for reclamation we have some information from archaeological studies. For example, it is now known that there was greater soil cover on Aran in the distant past, and that the thin soil cover seen today may have its origin in human interference, i.e. clearing land of small trees and shrubs for agricultural purposes and subsequent erosion of soil. Paleobotanical evidence from the Burren suggests that parts of it had a cover of pine, with subsidiary hazel and yew.[30] The notion that this also occurred in Aran is suggested by the place name Eochaill, one of the townlands of Árainn, *eo* meaning 'yew' in Old Irish.

The presence of the great forts would suggest that there was greater fertility of land on Aran in the past than today. Nonetheless, it is likely that the hinterland of these great forts stretched to the mainland of County Clare and the western coastline. These are not situated at the base of the inland cliffs, which we are accustomed to associate with ledges of fertile land supported by the shale layer of the terrace and protected from erosion by the scarp. The Celtic lords had sufficient resources to provide their own shelter in the form of massive drystone walls built as a duty by their clansmen and followers.[31] They are usually situated in a prominent position, slightly in the lee of the highest ridge of the island. The 1839 Ordnance Survey map shows only patches of good land around the cashels; they are not surrounded by good land in these maps.

Tom O'Flaherty, writing in the early twentieth century, says that the cashels or forts were situated on the best land in Aran. He describes the land around Dún Eoghanacht as if he were viewing it from a liner returning to Aran from America:

> We pass Creig a Chéirín. The cliffs grow taller. We are opposite Eoghanacht, the site of one of the four forts in Aran Mór. Archaeologists are in doubt about the origin of those fortresses, when they were built or by whom. But I came to the conclusion that the builders had a keen appreciation of the value of land, for this and the other forts in the island command the choicest pieces of land in Aran Mór. From the tall cliffs on the south of Eoghanacht to the sloping beach on the north there is a fairly decent pasturage, water is abundant, there are some large fields on which the soil is from four to six inches deep while the shore on the Conamara side is one of the best in the island for seaweed.[32]

The details of the Strafford Survey recorded in the *Books of Survey and Distribution*[33] gives us an estimated amount of 'arable and good pasture' and an amount of rocky land in each townland on the three islands of Aran in 1632. Adjusting their estimates to percentages, the townland of Cill Éinne had 56 per cent rocky land, Eochaill had 65 per cent, Cill Mhuirbhigh had 57 per cent and Eoghanacht had 64 per cent. On Inis Meáin, 57 per cent of the land was rocky, and on Inis Oírr 64 per cent. So at that time only a little more than a third of the land was usable for farming. Even if the maps of the Strafford Survey still existed, they may not have indicated where the different qualities of land occurred. Hence, we cannot be sure about where this usable land was located, but we can make some educated guesses. Firstly, it was spread across the islands because an amount of usable land is given for each townland. It is also likely to have been close to settlement locations. The figures may not be reliable as the Strafford surveyors, while confident about their acreage figures, admitted to uncertainty about their judgement of the quality of land.[34]

Roderic O'Flaherty's description of agriculture in the Aran islands in his *Chorographical Account of West or hIar Connaught* of 1684 has a more positive air:

> Beefe, butter, tallow, hides and of late cheese out of the Isles of Aran . . . The soile is almost paved over with stones, soe as, in some places, nothing is to be seen but large stones with wide openings between them, where cattle break their legs. Scarce any other stones there but limestones, and marble fit for tomb-stones, chimney mantle trees, and high crosses. Among these stones is very sweet pasture, so that beefe, veal and mutton are better and earlier in season here than elsewhere, and of late there is plentiful of cheese and tillage mucking and corn is the same with the seaside tract. In some places the plough goes.

This description occurs fifty years after the Strafford Survey. There is no mention of potatoes, but there is tillage, corn is being grown and there is cheese making. Cheese and various 'whitemeat' products belong to the era of Gaelic farming and ceased with the loss of the traditional infrastructure.[35] Roderic O'Flaherty describes a pleasant agricultural scene, which seems in contrast to what we know of the rest of Ireland in the seventeenth century. Agriculture in particular was being disrupted by the wars and the displacement that followed the Ulster plantation. Following the Rebellion of 1641 and the Cromwellian assault, crops were systematically destroyed, animals slaughtered, and whole territories were laid waste.[36] Cromwellians went to Aran and a garrison was stationed at the castle of Arkyn near the present-day village of Cill Éinne, formerly a 'Mannor and Castle' of Elizabeth I, until after the Cromwellian wars.

Today, if we are to judge by the CORINE (Coordination of Information on the Environment) land-use diagram, which shows land cover and land use, approximately nine-tenths of the land on Árainn is usable farmland, two-thirds is similar on Inis Meáin and all of the land on Inis Oírr is usable. We can deduce from this that the standards by which land is judged have changed in the intervening centuries or large tracts of land were reclaimed from the rock, using seaweed and sand and the craft of 'making land', passed down from father to son through the generations.

We must turn now to the question of why the fertile land was also subdivided with such intensity. Various examples of this can be seen when comparing the revised (1899) and the first edition (1839) Ordnance Survey maps. A vivid example is the area around An Caiseal in Gort na gCapall. These fields are clearly shown as good, fertile land in 1839, and they are likely to have been so, given that they occur on the shale terraces quite near the village settlement. By 1899 each field is divided at least in two, some in four. The fertile fields were subdivided, one supposes, to farm the existing pasture more intensively, under the pressure of population expansion and economic hardship. It may be that these areas were being used for crops and required stone walls for shelter. Did the stones come from other unreclaimed areas? It seems unlikely that these areas would have been described as 'good arable pasture' in 1839 if they were strewn with stones. It may be that communal farming practices that belonged to the Gaelic system continued up to this period. Or it may indicate a return to communal farming in times of hardship. The practice of partible inheritance or the Irish version of gavelkind, though outlawed, was still practised in Ireland and may also have played a part in this subdivision. A description of farming in Westmeath in 1682 describes the division of fields for tillage into smaller and smaller units:[37] 'They divide usually one field into acres, half-acres, stangs, that is roods; and of these they make so many lots or equal shares, as there are ploughs in the town; so as a man whose share may amount to three acres, shall not have perhaps half an acre together, but scattered up and down in all quarters of the field.'

While our contemporary sense of logic can grasp the practice of reclamation of land from the rock, clearing the fields of stones and building walls for protection of crops, the division of the fertile land is likely to remain a riddle, out of reach of our current thinking. It may belong to an earlier way of farming and responding to the exigencies of time and space in a different societal framework.

Relations between the tenants and their immediate overlords, the middleman James O'Flaherty, son of Patrick, and the agent Thompson, deteriorated in the second half of the century[38] during this period of frenetic wall building. The boycotts and protests of the Land Wars forced the government to introduce various Land Acts towards the end of the century, which culminated in the eventual transfer of the land into the ownership of the former tenants after the War of Independence in 1922.

FIGURE 27 Ceathrú of Eochaill showing reclaimed land today.

While a primary influence of the cultural landscape is geological, which is seen with particular clarity on the islands, the pattern of settlement and agriculture based on the land-unit matrix of the Gaelic era and overlaid by the landlord and tenant system of the eighteenth and nineteenth centuries is also expressed with a manifest clarity. The densely subdivided fields enclosed by drystone walls, which reached their climax of division and subdivision at the end of the nineteenth century, record among other things the adjustment from one political and economic system to another. These intensely divided and minutely constructed fields reflect the land hunger and hardship of that period, as well as the ingenuity and dedication of the island communities. For reasons of heritage and the record of social history in the landscape, we are fortunate that Aran, for whatever reason, was not 'improved' during the nineteenth century. Even the Congested Districts Board concentrated its efforts on the fishing industry, and left the farms unconsolidated. Some efforts towards consolidation of holdings occurred in the twentieth century, carried out by the Land Commission, but in general the pattern of landholding remained. These landscapes retain a remarkable continuity with the distant past, in particular with regard to the land-division system where ancient boundaries still exist. Even today, one can trace the relationship between a house in the village and groups of fields dispersed throughout the *ceathrú*, allowing each family access to the various types of land within the self-contained economic unit. In this way the pattern expresses vestigial forms of medieval farming as well as the later response of the people to the exigencies of time and place. The cartographic and documentary evidence leads one to conclude that the landscape slowly transformed from a moderately enclosed one to a complete web of enclosure that stretched across almost the whole of the islands, as each generation reclaimed the rock and subdivided the existing farmland.

CHAPTER 4

A Closer Look

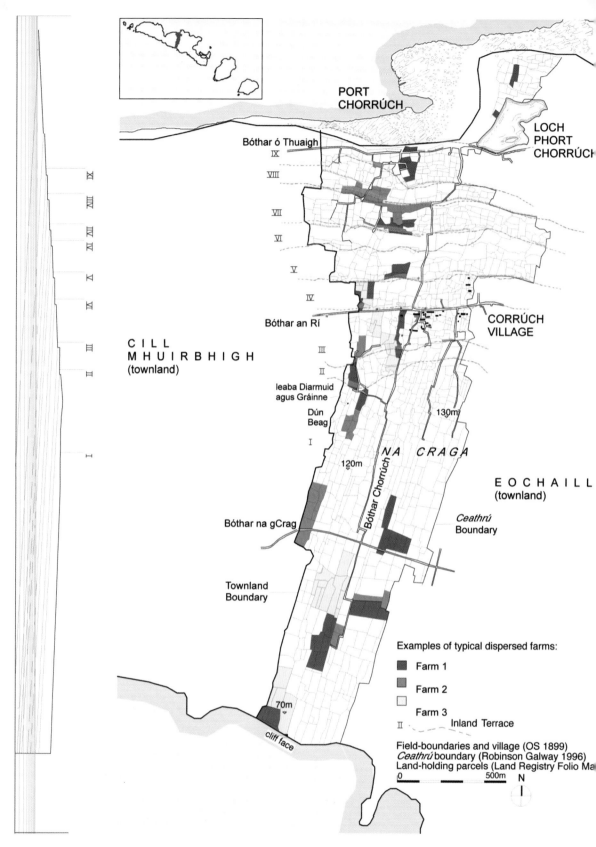

FIGURE 28 Map and diagrammatic section through Corrúch.

Ceathrú Chorrúch

The *ceathrú* of Corrúch is situated in the middle of Árainn in the townland of Eochaill and contains approximately 370 statute acres. One could say that it is a typical *ceathrú* . The boundaries begin high on the southwest coast and cut across the island, parallel to the geological sets of joints on the N10°E direction. It is bounded to the west by the more fertile and untypical *ceathrú* of Gort na gCapall, and this boundary coincides with the townland division between Cill Mhuirbhigh and Eochaill. The eastern neighbour is the less well-endowed *ceathrú* Baile na Creige. The latter seems an unusually narrow *ceathrú* but to compensate, it contains almost all of Loch Phort Chorrúch, the freshwater lake at the northern coast.

The highest point, 121 m (400 feet) above sea level, is on the uppermost terrace (terrace no. 1), some 90 m (300 feet) due east of the wedge tomb known as *Leaba Diarmuid agus Gráinne* ('the bed of Diarmuid and Gráinne', legendary lovers from an Irish mythic tale). From here, the sparsely covered limestone pavement, about 370 m across, slopes gently to the cliff face, where the land drops dramatically to the sea 70 m (234 feet) below. This point is named Corker on the Ordnance Survey maps, the anglicised version of *carcair*, an Irish word meaning 'slope', and used commonly to refer to the numerous slopes of this type to be found throughout the islands.

This stretch of land from the wedge tomb to the cliff face is Corrúch's share of Na Craga, which is used for wintering cattle. Wintering on the crags is an unusual variant of the ancient practice of transhumance. Elsewhere, farmers move animals to the uplands in summer and keep them on the lowland areas during winter. However, on the islands and also in the Burren on the mainland in County Clare, cattle are wintered on the upland terraces. This area of craggy limestone pavement takes up about two-thirds of the land in the overall *ceathrú*. Due to the oceanic climate, the winters are mild; cattle can be wintered outside without even being fed hay, as they would be elsewhere in Ireland. Brendan Dunford, in his book *Farming in the Burren*, describes the effect of the limestone base underlying the sparse grassland: 'The limestone of the Burren hills is said to act like a giant storage heater, building up heat in summer and dissipating it in the winter months when the surrounding atmosphere is cooler. As a result, a relatively warm, dry lie is ensured for outwintering livestock, with the thin sparse soils scarcely masking the limestone's warmth, while also proving very resistant to waterlogging, muddying and erosion.'[1]

Here on the crags is also where the spectacular diversity of unusual wild flowers grow – for which the islands as well as the Burren are famous. Geology, climate and agricultural practice combine to create the conditions where rare species of flower can emerge and proliferate. The drystone walls of the enclosed fields and the fissures of the limestone pavements provide the shelter necessary for certain rare plants to flourish. The purple milk vetch, *Astragalus danicus*, is one such plant, which does not occur in the rest of Ireland and is usually found in areas of a more continental climate.[2] Others – such as the gentian, hoary rock rose, the Salzburg eyebright and the maidenhair fern – abound. The absence of cattle in these areas during the spring and summer and their presence in the winter months is thought to be an important factor for the continuance of these species. The rendzina soils of the crags are nutrient-poor and this, combined with the winter-grazing regime, prevents the more competitive plant species from taking over the vegetation. The herb flora can then prosper and the tiny flowers bloom in spring and summer when the cattle are absent, grazing on the summer pastures of the lower terraces. The indentations of animal hooves are also thought to provide suitable niches for seedlings, and of course manuring by animals further enhances the habitat for these unusual plants.[3]

Just north of the wedge tomb, the terraces begin to step down towards the northern coast. These terraces support some of the limited indigenous fertile soil of the *ceathrú*. Terraces 3 and 4 are sprinkled with wells and, following the pattern described earlier, this is where human settlement occurs. The village of Corrúch is situated on terrace no. 3 close to a group of wells. It is a village of about seventeen houses. Of these, six or seven are dilapidated and no longer inhabited. It still retains its clustered village form. Just one house has been built very recently to the north of the *ceathrú*; it stands on its own more than half a kilometre from the village. To the southwest of the village is Teampall an Cheathrair Álainn ('the church of the four beautiful ones'), a ruined fifteenth-century stone church, encircled by three wells. This used to be a place of pilgrimage on 15 August, the Feast of the Assumption, each year and the holy well is reputed to have a cure for blindness. Some local people still make the pilgrimage. It was here that Synge set his now-famous play *The Well of the Saints*.

Evidence of earlier habitation is to be found on Terrace no. 2 and even at the uppermost terrace, in the form of collapsed clochans and the remains of other stone buildings. These may be part of the group of remains known as *Baile na mBocht* or *Baile na Sean* (village of the poor, or village of the old people), in the neighbouring *ceathrú* of Baile na Creige. This was thought to have been a medieval settlement, but archaeological excavation carried out in the 1950s was inconclusive.[4] To the southwest on the uppermost terrace and just on the other side of the townland boundary is An Dún Beag, the remains of an oval stone cashel of early date.

Terraces 4, 5, 6 and 7 had small enclosed fertile fields close to the base of the cliff on the Ordnance Survey of 1839, the rest of the land being shown as rocky

FIGURE 29
Fields on the crags in Corrúch. The wedge tomb is left of centre on the land horizon.

pavement. Today, most of this land has been reclaimed and the limestone pavements are overlaid with enclosed fields of grass. The lowest terrace beside the coast is shown as sand in the nineteenth-century Geological Survey. The 1839 Ordnance Survey shows enclosed fields of fertile land in these parts of the *ceathrú*. These field boundaries did not change very much in the second half of the nineteenth-century.

The *ceathrú* boundary includes the western shore of Loch Phort Chorrúch. Some of the holdings include a piece of shore from which water can be drawn in dry weather, or animals watered. In the past people fished for eels here. The lake is a good example of a karstic lagoon with cobble barrier, a type that is believed to be rare in Europe, and supports some rare lagoonal flora and fauna.[5] At present, it is also home to a family of swans.

There is some difference of opinion on the derivation of the name Corrúch, but one possibility is that it derives from *corr fhuach*, meaning 'bent or uneven cove', and refers to the coastal outline of Port Chorrúch, the natural harbour of the northern coast. This harbour was a great advantage to the *ceathrú*. Tomás Ó Fátharta (born 1922) of Eochaill described it as a marvellous harbour for seaweed 'stretching out to the broad Atlantic'. He recalled the place as a hive of activity, with people constantly coming and going with horses and donkeys carrying baskets of seaweed to create land and to manure existing land. The seaweed was brought in with the tide and was gathered when the tide went out. The men of the village often had to stay by the shore at night waiting for the tide to recede. The women came down with tea to keep them awake. Seaweed was harvested in spring for use as fertiliser on the land in November. Seaweed for kelp was harvested in early summer. The ruins of a large stone factory, built in the late nineteenth-century for making fertiliser from seaweed, stands at the harbour. Tomás recalls coming home from school along the lower road when the children could hardly keep their eyes open, the air being so full of smoke from the burning seaweed.

His family also had land in the *ceathrú* of Eochaill, but they availed of their right to use Corrúch because of the harbour there. The other ingredient for making land was sand, drawn from the beach at Cill Mhuirbhigh. This sand seemed to be available to everyone, even though it was in the *ceathrú* belonging to Kilmurvey House. Tomás recalls a tremendous amount of sand there, much more than there is today. (As well as being depleted by farmers for manure, it is also blown away by the wind.)

Fields near the coast were reclaimed first because it was a shorter distance to draw the materials. However, seaweed and sand were also drawn up the hill to the village of Corrúch and to the fields of Na Craga beyond. The horses used to be 'black with sweat' from the labour. The fields of the lower terraces of Corrúch often have larger than usual 'gaps' filled up with stones. This type of gap is called *bearna gcapall* because it is wide enough to take a horse or donkey with creels of seaweed on either side. The work of 'making land' was hard labour and it happened slowly. It has been estimated that the average Aran family reclaimed an acre of land for crops and some for grass over two generations. We cannot say when this method of reclamation began, but we know that it continued into the twentieth century.

John Millington Synge gave an account of land-making when writing about Inis Meáin. He had spent some weeks on the island each year from 1898 to 1902.

The other day the men of this house made a new field. There was a slight bank of earth under the wall of the yard, and another in the corner of the cabbage garden. The old man and his eldest son dug out the clay, with the care of men working in a goldmine, and Michael packed it in panniers – there are no wheeled vehicles on this island – for transport to a flat rock in a sheltered corner of their holding, where it was mixed with sand and seaweed and spread out in a layer upon the stone.[6]

FIGURE 30 The harbour and lake at the northern coast, Port Chorrúch and Loch Phort Chorrúch.

Dara Beag Ó Fátharta (b. 1922) of Inis Meáin gave an account of the process, as follows:[7] firstly, grykes (fissures) in the limestone pavements were filled in with little stones; plants growing in the grykes were left there because these were important in giving the new soil mixture a grip; this was known as *buan*, a very important element in the process – it could not be done without the *buan*; the seaweed and sand was mixed and laid in layers; many layers had to be laid down in order for the system to create soil; a lot of bulk was necessary. *Buan* refers to the indigenous soil, what had always been there, and it was paramount to have some small amount of indigenous soil. This is why the islanders in Synge's description above treated the small amount of soil in the corner of the cabbage garden as if it was gold.

The pattern of landholding is one of unconsolidated holdings where each household of the village had land dispersed across the *ceathrú*. In order to farm the *ceathrú*, each farmer required some of the different types of land that the unit offered. As described above, this system is inherited from earlier times, based on the viable economic unit of the land-division system. Gradually every plot became individualised, while some of the work was still carried out in common until the late twentieth century. This self-sustaining unit, which offered the possibility of diversified agriculture, contributed to the survival of the people through the famine eras, in contrast to the nineteenth-century clachan and rundale settlements of the Connemara coast where communities were farming areas of wasteland of mountain or bog, which had never been self-contained sustainable units, with tragic consequences.

Today the pattern of dispersed landholding persists, although families now often own land in various *ceathrúna*, through inheritance or through purchase.

In Corrúch today there are about 19 holdings over 10 acres; the largest holding is 33 acres. In 1917, the Land Commission carried out a survey of rented lands on the islands and its Schedule of Areas lists the sizes of farms in acres, the acreage for each parcel of land, and the names of the tenants. At that time, about 250 acres were being farmed by people who lived in the village of Corrúch; the other holdings were being leased by people from neighbouring villages. At that time there were just 19 acres of commonage in the *ceathrú*. These acres have now become part of individual holdings. The holdings have changed little since that date. In general, the holding size is the same. The same surnames occur in the village then as now: Ó Fátharta; Ó Donnchadha; Ó Conghaola; Ó Concheanainn, and Ó Dioráin, among others.

Often, dispersed holdings in other parts of Ireland were consolidated during the nineteenth and early twentieth century (the classic example, complete with before and after surveys, being the one in Clare Island, off the coast of County Mayo). One of the reasons why this did not occur in Aran may be that there really was no other way to farm the land. If it had been stripped from coast to coast, the strips would have been unworkably narrow, given the small size of the holdings. Elsewhere, consolidation of holdings was at times met with opposition from tenants. They preferred to have pieces of land in different parts of the *ceathrú*. However, the opposite was also true. When the system became overly subdivided and unreasonably dispersed it lost its meaning and farmers felt aggrieved to have unworkably small parcels of land at remote distances from each other. It may be that in Aran the land-unit system and the dispersed farms within it was an appropriate response to the specific and contained nature of the island landscape and for that reason it endured. On the smaller islands in particular, there is a sense that each farmer farms the whole island, because each one has land in every part of it. Today, despite the almost complete individualisation of commonage and the decline in communal agricultural activities, the intermixing of holdings remains and continues to retain a greater communal dimension than a system of consolidated holdings.

There is a powerful sense of place in Corrúch, despite its labyrinthine quality on first experience. Understanding the system clarifies the labyrinth but does not detract from its mystery. One retains a sense of appreciation, if not awe, at the sheer extent of human endeavour, enhanced by the extraordinary setting on the edge of the vast Atlantic. The construction, in stone and by hand, of this landscape of small fields, walls, water troughs, stiles, wells and green roads that make up this village and *ceathrú* – a self-contained human habitat – is a record of its history.

Facing page: **FIGURE 31** Map of landholding pattern in Corrúch.

PORT CHORRÚCH

Bóthar ó Thuaigh

LOCH PHORT CHORRÚCH

Holdings of over 10 acres in Corrúch (2004)

*Holding 1 = 23 acres

ºHolding 2 = 22.5 acres

*Holding 4 = 14.5 acres

*Holding 5 = 17 acres

*Holding 6 = 13 acres

*Holding 7 = 16.5 acres

*Holding 8 = 15 acres

 Holding 9 = 6.5 acres

*Holding 10 = 11 acres

*Holding 11 = 20 acres

*Holding 12 = 10 acres

*Holding 13 = 32.5 acres

*Holding 14 = 20 acres

 Holding 17 = 21 acres

 Holding 18 = 17 acres

 Holding 30 = 14 acres

 Holding 41 = 10.5 acres

 Holding 49 = 18.5 acres

 Holding 70 = 20 acres

**Holding107 = 10 acres

*Landholder lived in village of Corrúch
 in 1917

ºOne brother lived in Corrúch, farm size
 has reduced since 1917

** Not recorded in 1917

(Information from Land Commission
Survey 1917)

CORRÚCH VILLAGE

Bóthar an Rí

townland boundary

ceathrú boundary

Bóthar na Craga

Field-boundaries and village (OS 1899)
Ceathrú boundary (Robinson Galway 1996)
Land-holding parcels (Land Registry Folio Map)
Terraces (GSI 1867)

0   500m   N

FIGURE 32 View north over Corrúch in summertime.

Facing page: FIGURE 33 Map of Ceathrú Chorrúch showing Study Farm.

FARM STUDY

The study farm is a 23 acre holding in twelve parcels in Corrúch, a *ceathrú* in the townland of Eochaill on Árainn. As with the other farms of the village the landholding is dispersed across the *ceathrú* from a one-acre field at sea level, bordering Port Chorrúch, the seaweed harbour on the north coast, to the most southerly field of the *ceathrú* at the Atlantic cliffs, 73 m above sea level. At the time of the granting of this farm by the Land Commission in 1927 the house was in the village of Corrúch, perhaps 45 m from their nearest field. Later a house was built in one of the fields at the western edge of the village. The fields around the village comprise in total about 3¹/₂ acres in four parcels. These fields are used as pasture for cows and calves, and some goats. The two half-acre fields near the townland boundary are set in meadow for the making of hay.

The parcels to the northwest of the village, were in general formed at the time of the 1839 Ordnance Survey. They are part of the arable land in enclosed fields around the village, except for the most northerly portion on the lower crag which has been reclaimed since the first edition Ordnance Survey. This was part of the reclamation carried out during the second half of the nineteenth century. It has been estimated that the average Aran family reclaimed an acre of land for crops and some for grass over two generations.

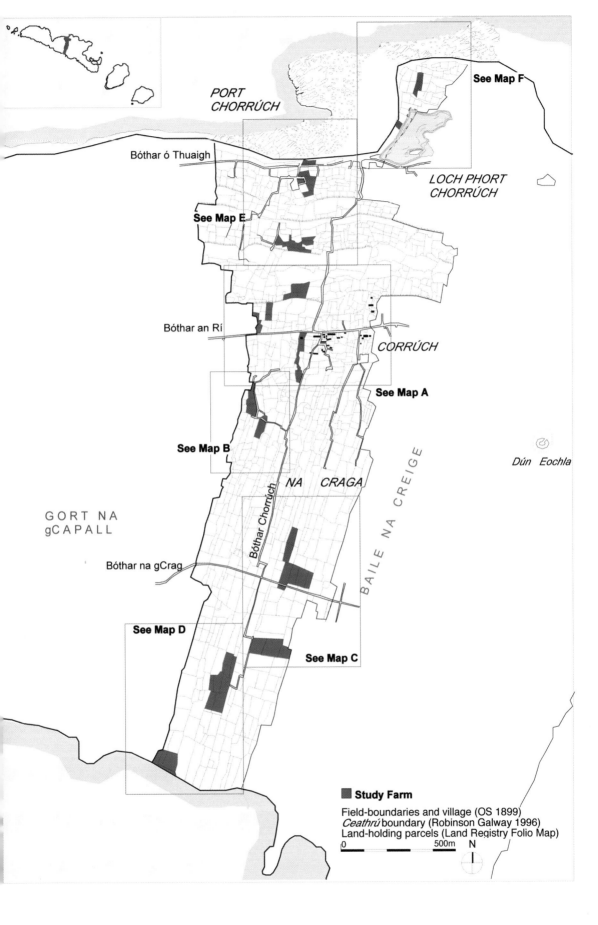

PORT CHORRÚCH

See Map F

Bóthar ó Thuaigh

LOCH PHORT CHORRÚCH

See Map E

Bóthar an Rí

CORRÚCH

See Map A

See Map B

Dún Eochla

NA CRAGA

GORT NA gCAPALL

Bóthar Chorrúch

BAILE NA CREIGE

Bóthar na gCrag

See Map D

See Map C

■ Study Farm

Field-boundaries and village (OS 1899)
Ceathrú boundary (Robinson Galway 1996)
Land-holding parcels (Land Registry Folio Map)

0   500m   N

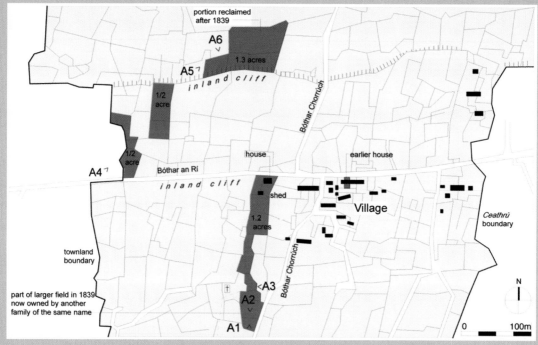

FIGURE 34 Map A: Fields around the house in the village.

FIGURE 35 A1 Looking north over the parcel of fields above the house. Note the open *bearna* (gap) to let cattle through.

FIGURE 36 A2 Looking south over fields behind the house, note the green road from Bóthar Chorrúch to the left leading to Na Craga, the winter pastures on the ridge of the island to the south. This field was part of a larger field in the 1839 Ordnance Survey, the rest of which is now owned by a family of the same name from the neighbouring village. The wall dividing the field is shown on the 1899 Ordnance Survey.

FIGURE 37 A3 Looking west over summer pasture fields. The ruin in the middle ground is the fifteenth-century church Teampall an Cheathrair Álainn. In the middle distance is Port Mhuirbhigh, the harbour at Kilmurvey.

FIGURE 38 A4 These two half-acres at the edge of the townland boundary are clearly being managed by the same farmer. Note the scrub-filled field in the foreground on this side of the townland wall, and the erratic in the field beyond. These fields were described as arable in the 1839 Ordnance Survey.

FIGURE 39 A5 The field outlined in red was probably reclaimed from the limestone pavement after 1839, although it remains a rocky field. Fields in the foreground are part of the indigenous fertile ground of the shale layer. The new 'one-off' house built about 500 m from the village of Corrúch can be seen in the distance.

FIGURE 40 Left: A6 A spring emerging from the base of the inland cliff that forms the southern boundary of this group of fields. The photograph was taken after a period of heavy rain in winter. Right: The spring viewed from above.

Further south and uphill towards the crags two parcels of land are connected by an access road. These in total comprise 1¾ acres. The more northerly field falls steeply towards the north and has access at the bottom to a well. These fields are enclosed in a similar way in the 1839 Ordnance Survey and the access roads are present. Access to the well rendered the land usable before the stone water-collection troughs were built. A wall shown in the centre of this field in the 1899 Ordnance Survey is now gone. The field to the south is part of Na Craga, which is used for wintering cattle. This is where many stone water troughs are to be seen, because here on the upland region there are virtually no wells. There are two walls here, built since 1899.

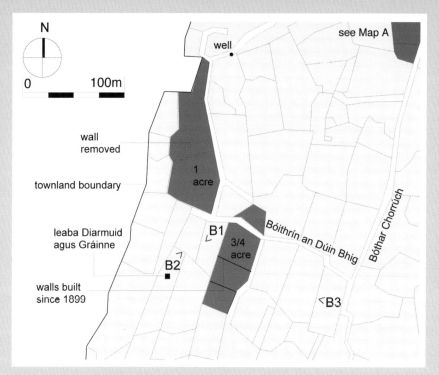

FIGURE 41 Map B: Study Farm in Corrúch showing fields uphill from the house towards the crags.

FIGURE 42 B1 Looking southwest over Na Craga: in the foreground the sparse layer of soil barely covers the limestone pavement. In the centre ground is the wedge tomb, Leaba Diarmuid agus Gráinne (note the stone water trough in front of the monument).

FIGURE 43 B2 Looking north, the red dots mark Bóithrín an Dúin Bhig; the second access road leads down to a well. This road would have been very important to farmers for watering cattle in the past. Now the lower reaches of it are impassable because scrub has taken over. In the foreground, right, is the back of a stone water trough. The hills of Connemara are on the horizon.

FIGURE 44 B3 The stone rainwater-collection troughs of Na Craga. There is usually one for each group of fields. There is no other source of water for animals in these fields. In the lower parts of the islands the troughs are sometimes filled from wells in summertime. Here, they are completely dependent on rainwater. This is another reason why the crags are suitable for wintering animals: there is likely to be more rain, and the animals are less thirsty in cooler weather.

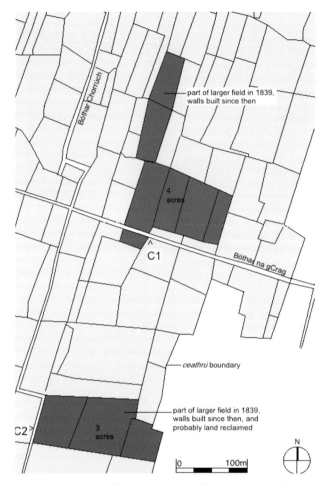

FIGURE 45 Map C: Fields in Na Craga used for wintering cattle.

Further south towards the middle part of the crags, two parcels of land have fields that are somewhat bigger than in the lower terraces to the north. Fields are bigger where the grass is scarce. In general, the fields are dry in winter, which is another feature that makes them attractive for winterage. Nonetheless, flooding can occur after very heavy rain even up here on the first terrace; however, cattle would be able to find shelter and a dry lie. While there are often fierce storms with high wind and driving rain, the mild oceanic climate precludes frost or snow.

Bóthar na gCrag was built in the late nineteenth century. Farmers had to give up some of their land for the new road: sometimes parcels of land were split to allow the road through, as has happened in this case.

FIGURE 46 C1 Looking north over the four-acre parcel of fields. This picture was taken after a particularly wet period in January. This field was shown as fertile land in the 1839 Ordnance Survey. The fields beyond the stone wall on the horizon were described as rocky.

FIGURE 47 C2 Looking east over fields to the south of the map. These walls have been built since the 1839 Ordnance Survey. This land is likely to have been reclaimed since; however, the location is about 2 km from the northern shore from where the seaweed and sand would have been drawn by a horse with creels. The limestone pavement can be seen above the soil cover in places; this is the kind of land that is described as rocky in the Ordnance Survey maps.

At the southernmost part of the crags there are two further parcels of land. The larger parcel is accessed at the very end of Bóthar Chorrúch. Access to the furthermost field at the cliff edge is by right-of-way; it is likely that access is by agreement between neighbours who may also be relatives. The larger parcel is a set of flat fields where the geological jointing is very close. In between the fissures, grass accumulates. In the northern parcel, none of the boundaries has changed in the 100-year interval since the 1899 map.

The smaller parcel is right at the cliff edge where the land drops 73 m (240 feet) to the sea below. Here strong east–west joints in the limestone bed have created a cliff face which is constantly being eroded by the sea. The parcel has now just two fields. The name Corker (*carcair* meaning 'slope'), which appears on the Ordnance Survey maps, may have been given to it because of the slope shown below. Its extraordinary location on the cliff edge means that it is constantly sprayed with a vaporous sea mist that has risen from the crashing waves below.

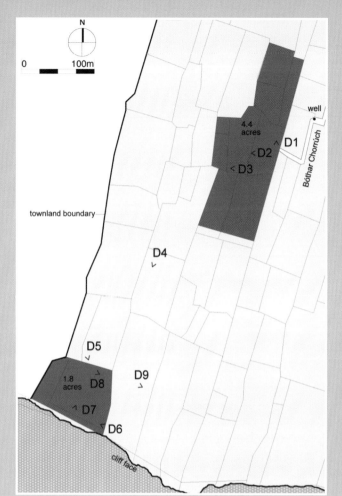

FIGURE 48 Map D: Study Farm in Corrúch showing fields on the uplands of Na Craga.

FIGURE 49 D1 Looking north over a larger parcel of fields. The 'ridges' in the field parallel with the field boundary are the fissures in the limestone pavement.

FIGURE 50 D2 Looking west over the same field. Note the deteriorating drystone wall.

FIGURE 51 D3 Looking west. The silhouette of Dún Aonghasa is on the horizon, middle left.

FIGURE 52 D4 Stone water trough used to store rainwater for cattle, generally the only source of water on the crags.

Often there is no source of drinking water on the crags, but here there is a well nearby, at the end of Bóthar Chorrúch, known as 'the well of the still-house.' In the 1839 Ordnance Survey, there is a small building shown; no trace of it remains. Probably it was a poitín still once. (A well was required for steeping the barley-grain to make whiskey.)

FIGURE 53 D5 Looking southeast over the last field of the farm at the south coast. Note the stone water trough built into the slope of the field, and the exposed edge of the shallow limestone bed with the almost horizontal shale layer showing as a dark line.

FIGURE 54 D6 Looking west over the last wall of the farm on the southern coast. Note the shallow limestone bed jutting out where the shale layer has been eroded from below.

FIGURE 55 D7 Looking north across the last field. Note the fissures in the limestone pavement, set closely together.

FIGURE 56 D8 A stile of through stones (i.e. steps occur on the other side also, allowing access to this field).

FIGURE 57 D9 An unusual small trough in the neighbouring field, probably filled by hand from a larger one.

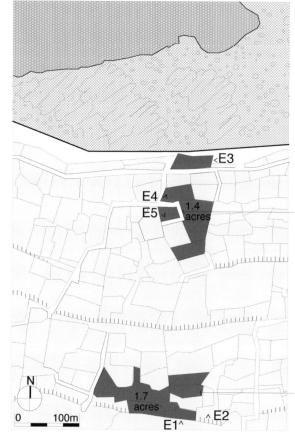

FIGURE 58 Map E: Study Farm in Corrúch showing fields on the lower terraces and at the coast.

What appears as a road here may have been a cattle pen. All this is gone now (see photos). A modern cattle pen has been constructed at the corner.

This is the field full of water in the photo. There has been a lot of change to boundaries in this parcel

FIGURE 59 E1 Looking north over terrace no. 6. A spring at bottom right is feeding the flooded field to the left.

North of the village there are two parcels of land; a group of numerous enclosed fields on the lower terraces, and a parcel at the coast. Reclamation and enclosure has occurred here since the 1839 Ordnance Survey and boundaries have also been altered since the 1899 Ordnance Survey. Manuring and enlarging of fields has obviously occurred in the twentieth century. In the parcel to the south, boundaries have changed so much that it is no longer possible to trace the outline of the parcel in the landscape, in the way it has been possible with most of the other dispersed fields. These fields are dry in summer when they are used for pasture.

FIGURE 60 E2 A spring emerges from the limestone pavement beneath the layer of soil and grass after heavy rain and creates a stream that floods into the field to the left. The photograph was taken after a period of heavy rain in winter.

FIGURE 61 E3 Looking west over the field at the northern coast. This field existed in 1839. It was cut off from the other fields to the south when the road was built, to the left of picture. Being so close to the coast it has probably been manured and cleared over the years. At the far end of the field, middle left, is a pile of stones cleared from the field. Note the wire and steel post fence marking the boundary with the strand. This would have formerly been a drystone wall.

FIGURE 62 E4 Looking northeast over coastal fields on the lowest terrace. The photograph was taken in January after very heavy rainfall.

FIGURE 63 E5 Looking south over the same fields. These field-boundaries have remained the same except that the road shown on the map, which may have been a cattle pen, is gone and a new pen has been constructed in the corner of the field.

At the northern coast there are fields in three parts: one field bordering the lake which gives access to fresh water; one virtually in the lake during wet winter weather; and finally a group of sandy fields at the northernmost part. The access road shown to the north of the lake was built in the second half of the nineteenth century. At least half of the fields to the north of the lake were shown as rocky on the 1839 Ordnance Survey and must have been reclaimed later in the century. All are grassy fields clearly in use for grazing.

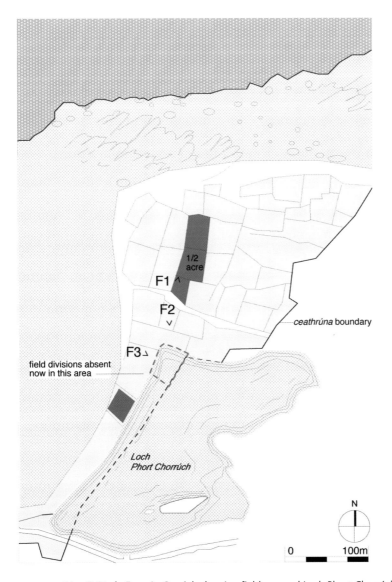

FIGURE 64 Map F: Study Farm in Corrúch showing fields around Loch Phort Chorrúch.

FIGURE 65 F1 Looking north over the sandy fields of the northern coast.

FIGURE 66 F2 Looking south over Loch Phort Chorrúch. The villages are on the higher terraces at the horizon: Eochaill on the left, Baile na Creige in the centre, Corrúch to the right. There are no longer divisions along the piece of strand to the right of picture; this seems to have returned to commonage.

FIGURE 67 F3 Looking east over Loch Phort Chorrúch: the red line shows the approximate postion of the submerged field.

WALLS

The purpose of building walls, apart from delineating boundaries between one farm and another, is to clear the fields of stones, to create enclosed fields for the purpose of intensive cattle rearing and to provide shelter for animals and crops. The system is flexible, durable, avoids the cumbersome and expensive business of making mortar, and fulfils the dual function of clearing the field and making an enclosure at the same time.

Most importantly, the drystone wall is constructed in such a way that it does not have to withstand wind pressure. This is crucial because most of these walls are subject to very high winds from the Atlantic. Unlike the thick (sometimes 2 m or even 3 m) drystone walls used for constructing forts or 'beehive huts' of earlier centuries, which depend on the weight and compression of many long stones piled on top of each other to withstand the force of Atlantic gales, the drystone field-boundary wall of single stones depends on friction between stones for its stability, and incorporates air spaces into its structure to let the wind through. While the former is massive enough to withstand the gales, the latter survives the onslaught of the wind by acting as a filter.

The sharp orthogonal stones of Aran are particularly good for achieving enough contact to create friction and yet have the shape and suitability to form a filtering screen. Seen against the sky, this lace-like tracery of myriad shapes and patterns, not one like the next, in the clear light reflected from the great Atlantic gives Aran its particular beauty and character.

FIGURE 68 The massive drystone walls of Dún Chonchúir, Inis Meáin.

FIGURE 69 Field-boundary wall in Eochaill, Árainn. The structural stability is dependent on friction between the stones. The wall acts as a filter for the wind.

FIGURE 70 A typical *fidín* wall, Inis Meáin.

SINGLE WALLS

The single drystone wall is one stone deep and generally about 4 feet high. It is constructed by clearing away an area of soil about 2 feet wide down to the rock (the soil will not be more than a few inches deep), then placing large vertical upright stones about 4 feet apart. These are known as *clocha máthar* (mother stones). In between the mother stones the wall is filled with small stones; on top of this foundation, stones are placed randomly. The lower level of stones is tightly packed, to strengthen the foundation, to stop rabbits getting through, and because close to the ground wind pressure is not an issue.[8] The lower part is usually about 2 feet high. The upper part has gaps between the stones to create the wind filter. This is the basic format for single drystone wall building, and it is the many variations of this basic layout that gives the various drystone walls their aesthetic appeal. This is also known in Aran as a *fidín* wall. However, not all walls display these characteristics. Patrick McAfee suggests that the *fidín* wall principle came from Scotland in the eighteenth century and was introduced to east Galway in the nineteenth century by improving landlords. He describes the wall as a combination double and single wall, the bottom being filled with small stones as double and the top having single stones.[9] This type of wall was insisted upon by the Department of Agriculture when it was grant-aiding wall building in the twentieth century.[10]

DOUBLE WALLS

Double walls were built to enclose roads: public roads and green roads. The islanders were prevailed upon to build the walls when the main road going through Árainn was widened in the nineteenth century. This wall had a stone on either side and a flat coping stone on top. Tomás Ó Fáthartha referred to this as a *balla dúbhailte* – 'double wall'. The word for coping was the same as the English word.

ANCIENT WALLS

As discussed in earlier chapters, the *ceathrúna* and townland boundaries are thought to be centuries old. These *claí teorainne* (mearing walls) were grant-aided by the Department of Agriculture for repair and maintenance during the twentieth century, hence it is likely that they were rebuilt in places over the last century. Certain men were good at wall building and were happy to avail of the opportunity to make some money when cash was scarce, before tourism became a source of supplementary income for farmers. The boundary is usually, but not always, marked by a natural feature such as a deep gryke or depression in

FIGURE 71 A double wall bordering Bóthar na gCrag, the road that traverses the crags on the southern ridge of Árainn. It was built as a Relief Works scheme introduced by Arthur Balfour (then the British Government's Chief Secretary for Ireland, and later Prime Minister) in the late nineteenth century.

FIGURE 72 A *fidín* wall in Inis Meáin with a low foundation or base. There were probably not many small stones in this field, and so the basic strategy was altered to suit circumstances.

FIGURE 73 A low *fidín* wall in Fearann an Choirce, Árainn. The beach at Cill Mhuirbhigh and the harbour, Port Mhuirbhigh, are in the middle distance; the silhouette of Dún Aonghasa on the horizon can be seen at the far left.

FIGURE 74 *Fidín* walls in Inis Meáin. Note the erratics used as a base.

FIGURE 75 *Fidín* walls in Gort na gCapall making use of round boulders – granite and sandstone erratics deposited in Aran from Connemara after the last Ice Age.

the land, or the edge of an inland cliff, so that often the position of the boundary was known even if the wall was falling down or deteriorating.

The question is often asked by visitors to the islands: why are the walls different in style from each other? The answer lies somewhere between the different styles of the men who made them and the terrain and type of stone particular to that place. For example, take the walls that border Bóthar Chorrúch: this green road is marked on the 1839 Ordnance Survey map, and recurs in the 1899 Ordnance Survey. The wall is remarkable for the very large stones of which it is made up. The limestone pavement is used as a foundation in a way that expresses very clearly the alignment of the walls with the geological cracking.

The building of such walls was likely to have been a communal activity since it is the wall bordering the access road used by all the farmers of the village. Probably a group of men from Corrúch built the wall using large stones. They clearly had some system for moving and lifting great stones, or they were powerfully strong men like the legendary brother of Big Seán Thaidhg from Connemara who could 'lift a two hundred weight stone by himself and throw it out of the way'. He had 'ribs like a boat and two shoulders on him like a rowing boat'.[11]

FIGURE 76 Boundary wall to the green road, Bóthar Chorrúch, looking north. The limestone pavement is used for the road, and the step in the pavement is the foundation for the wall.

FIGURE 77 Walls using the rock as a base, Árainn.

FIGURE 78 *Ceathrú* wall between Mainistir and Eochaill, looking north, in the townland of Eochaill. These walls are of particular interest because they mark an ancient boundary. It may be that parts of the walls are as ancient as the land division; other parts have been rebuilt over the centuries.

109

FIGURE 79 *Ceathrú* wall between Mainistir and Eochaill looking south. The Cliffs of Moher in County Clare are on the horizon to the left.

FIGURE 80 This wall in Cill Éinne townland was built in 1996 or 1997 – proof of a still-living craft of drystone wall building. It is in Ceathrú an Chnoic, the Hill Farm, originally attached to Killeany Lodge, which was divided up by the Land Commission in the 1920s and 1930s and distributed among farmers in the locality. Few walls appear in this *ceathrú* in the 1899 Ordnance Survey.

FIGURE 81 Walls in Ceathrú an Chnoic have been built since the division of the land but in general this *ceathrú* does not have the characteristic pattern of minutely divided fields common in the rest of the islands. On the horizon is the ruined Early Christian oratory known as Teampall Bheanáin.

FIGURE 82 New walls in Inis Meáin, clearly made with the help of machinery when the new road to the windmill farm on the west coast was being laid down.

FIGURE 83 A group of fields, Inis Meáin.

FIGURE 84 Walls in Eoghanacht made with lots of small stones characteristic of the area.

FIGURE 85 The structure to the right of picture is a wall closing off the scant but nutritious herbage of the limestone pavement, for later in the year.

Traditionally in the Aran villages certain men who were good at wall building would have availed of the government grants system to build walls and to help others in their villages. They became well known for their work and are still remembered. A wall builder from Eognanacht was such a person, and there the stones were very small, 'like sods of turf' laid diagonally on top of each other, expressing a particular pattern that is evident only in that area.

Field Furniture

BEARNAÍ, CÉIMEANNA, STAIGHRÍ

Until recently it was believed by visitors to the islands that the fields of Aran were completely enclosed by walls, with no specific opening, and that islanders knocked down some of the wall every time they wanted to enter the field.

One suspects that the error arose because the gaps and the stiles are quite difficult to see. They blend into the pattern of stones in the walls to the extent that the field appears entirely enclosed. However, to have fields without openings would not make sense, and Aran is a working landscape with its own clear logic. The islanders further confuse the matter by offering to 'knock down a bit of the wall for you' if they come across a hapless tourist trying to cross a stone wall and unable to find the stile. What the islander means is that he will knock the stones out of the gap (which he can of course easily see and knows where it is anyway since it is probably his field) and build it up again. The farmer is also trying to avoid having the tourist knock down a piece of the wall while crossing it, because it will have to be rebuilt, unlike the gap, which is emptied and filled with stones easily because it is designed for that.

Since Tim Robinson published his books and maps, it has become better known that there are gaps and stiles in most fields, and that these should be used in preference to crossing walls willy-nilly because this can damage the walls and also could be dangerous. He gives a lucid and amusing description of the Aran gap, written after a day wandering Na Craga when a rain shower had picked out the sparkling granite boulders filling the gaps against the rain-soaked limestone of the walls:

> This focused my mind on the question of gaps, and after some research I wrote the following little treatise:

> 'The Aran *bearna* or gap is no mere hole blocked with a loose assemblage of stones, but a specialised and adaptive structure. It is usually two or three feet wide, with an upright stone on either side, and often these jambs slant apart slightly so that the stones piled between them are held in the wedge-shaped space. The granite erratics brought over from Connemara by the last Ice Age strewn here and there on Aran's crags are preferred to limestone for filling the gaps, because they are naturally ovoid and very tough, so that the gap is easily "knocked" by tumbling the stones aside, and they do not crack up after repeated use. The Aran farmer and even his child can "raise up" such a gap in the time it would take an outsider to bruise his or her fingers arranging the first few stones of it. This temporary fence is unstable, and often a short length of briar or a blackthorn branch is wedged among its topmost stones to discourage cattle or horses from nosing it down. Clearly, the gap is not the place to climb the wall since it is built for collapsibility but somewhere close to it will be a stile, or at least a through stone or two adequate to the practised foot of the landowner.'

> Having thus formulated the Aran gap, I wandered out to have another look, eyes sharpened by theory. And behold! Every conceivable ad hoc concoction of concrete blocks, thorn bushes, driftwood, worm-eaten oars, carcasses of oil drums, iron bedheads, complicated pipework looted from wrecks,

FIGURE 86 A gap and stile nearby to the left with shoes on the through stones.

bicycle-frames, anything and every thing redundant and outworn will serve to stop a gap just as well as the granite boulder.[12]

Like Aran walls, Aran gaps defy categorisation, however, there are many gaps that fit the theory.

Each group of fields owned by the same farmer will have gaps connecting the fields so that they can be opened or closed depending on how they are being grazed. Usually the field bordering the access road will have a *staighre* (stile) nearby for the farmer to cross with ease. The stile is made by putting a long stone horizontally through the wall at intervals to suit stepping over the wall – there is no need to break down the gap for human beings: the gap is for moving animals through the fields.

There are also *bearna gcapall*, gaps wide enough for a horse or donkey with creels of seaweed to pass through. There are many of these to be seen near the seaweed harbours where land was reclaimed in the nineteenth and twentieth centuries.

FIGURE 88 *Bearna gcapall* near Ceann Gainimh, the sandy beach in Inis Meáin.

FIGURE 89 *Bearna gcapall* near Port Chorrúch (a seaweed harbour), Árainn.

Facing page: FIGURE 87 A gap leading from the green access road in Corrúch to a set of fields, stretching northwards. This rather beautiful gap comes close to Robinson's platonic ideal.

FIGURE 90 A gap between fields in Inis Meáin, with a stile to the right. Here the wedge stones of the gap are added to, in order to make a kind of stone pier. This is quite common in Inis Meáin where the walls are higher than in Árainn.

FIGURE 91 A gap from an access road to a field, Inis Meáin. Through the top of the gap a stone water trough can be seen in the field.

FIGURE 92 A stile for accessing a field from a green road in Eochaill, Árainn.

FIGURE 93 An adjacent gap for animal access to the same field.

FIGURE 94 A stile at an access road in Eochaill with a prominent through stone.

There are two kinds of stile. One is known as *staighre*, which consists of stones going through the wall forming steps. The other, *céimeanna*, is where there is an opening at the top of the wall, similar to what is normally called a stile, which designates a right of way. The *staighre* occurs within a farmer's own set of fields or from the access road to the field. *Céimeanna* occur, for example, at the entrance to a well or on a route through fields to a monument, and also indicating short cuts and routes to the shore. There is intriguing folklore written by Robinson and others about these 'secret' paths.[13]

FIGURE 95 Tomás Ó Fátharta of Eochaill standing at the stile entrance to Tobar Nora, in the far north of Ceathrú Chorrúch.

FIGURE 96 Tobar Nora. The well is reached through this stile from the access road. Wells like this one, which is used by all the villagers, usually have a kind of forecourt or garden and stile entrance. Tomás Ó Fátharta pointed out 'the fine flat stone' at the stile 'for placing your water container on'.

FIGURE 97 A stile in Inis Meáin.

FIGURE 98 A gap at the high point of the island in Árainn looking northwards to the Connemara coast. In the foreground left is a stone water trough for watering cattle.

WATER TROUGHS

A method of rainwater storage was developed in Aran for providing animals with water. This was particularly necessary on the upland areas of the crags where there are few natural wells or other sources of fresh water. The standard rainwater tank has a rectangular trough, *an iomar* ('the font'), which collects the water, and a slanted part built up with stones and finished with flat flags, *an teachín* ('little house'), which directs water into the trough. The flat flags on the slanted roof of the *teachín* are sometimes coated with a cement mixture to provide a smooth surface. The trough is usually about 1.8 m long, 0.9 m wide and about 0.6 m high, the height required for cattle to drink from it. There are many variations on this theme; some are quite large, more than 3 m long. Among the islanders there are various theories about the origin of the stone rainwater-collection troughs, which now proliferate throughout the islands, but are most plentiful on the crags. One theory is that they were invented by Roger Dirrane from Cill Éinne, who lived in the twentieth century. Older farmers recall a time when there were no water storage troughs and how difficult it was in times of drought to bring water to cattle. Dara Beag Ó Fátharta remembers making a stone trough in the 1950s and having to wait for weeks before being able to buy a bag of cement in order to make it waterproof.

Sometimes there are sloped structures on two or three sides to increase the amount of water reaching the trough. There is also a version with two troughs. Troughs are often built into sloping ground to employ economy of means.

FIGURE 99 A typical stone trough.

FIGURE 100 This trough on the crags in Eochaill is 4.84 m long. The main stone forming the trough is 3.7 m long and is cemented to another to extend the length further. The trough part is 1.33 m wide and 1 m deep, the *teachín* is 2.3 m x 4.84 m. It has a smaller trough to one side, about 0.48 m high, from which cattle can drink.

FIGURE 101 The trough is too deep for cattle to drink from, so a smaller one to the side is filled by a copper pipe and tap that connect to the larger one.

FIGURE 102 See the smaller tank to the bottom left of the picture.

FIGURE 103 A double trough in Inis Meáin with three sloped parts, built into the slope of the land. The lower trough can be reached by cattle.

FIGURE 104 A double-sided trough on flat ground at the northern end of the *ceathrú* Eochaill in Árainn. The village of Eochaill can be seen in the distance on the horizon.

FIGURE 105 A trough on the ridge of Árainn at Eochaill. Advantage is taken of the fall in the land, to minimise the amount of building for the sloped part.

FIGURE 106 A closer look at the same trough.

FIGURE 107 A small trough, Corrúch. This particularly beautiful small water trough is just inside the wall of Bóthar Chorrúch at the crest of the ridge above the village. The walled access road – Bóithrín an Dúin Bhig – can be seen crossing the fields beyond.

FIGURE 108 A combined shed and trough in Inis Meáin. The shed is used for storing potatoes and crops.

FIGURE 109 An Tobar Mór ('the great well'), a natural collection point at the base of the inland cliff for the distilled water that has travelled down through the grykes of the limestone pavement and along the impervious shale layer.

FIGURE 110 An Tobar Mór was renowned throughout Árainn as a well that never ran dry. It is situated at the bend of the road in Fearann an Choirce, about halfway across the length of the island, and was used in times of drought for drinking water even after piped water was installed. The outer wall is man-made to contain the water because it was such a reliable source.

Bóithre (roads)

Aran has a network of roads which facilitated farmers' access to their dispersed holdings across the islands. The roads also gave access to the harbours for the collection of seaweed and kelp making, and from which to launch boats for fishing. There are many green roads within the agricultural landscape; they are always walled. Some have become 'real' roads with more recent changes such as Bóthar an Aerstruip ('the airstrip road') in Inis Meáin, formerly known as An Seanbhóthar ('the old road'). This road existed in 1839, and was clearly very important to the islanders because it led to Ceann Gainimh, ('the Sand Head'), from which sand could be drawn for reclaiming land. It also gave access to the natural harbour at An Caladh Mór. Nowadays, it goes to the small airport, a new pier and the marvellous sandy beach in summer at Ceann Gainimh. Already many access roads, particularly the smaller remote ones, have become impassable through disuse. Human and animal passage on these roads is what preserves them from becoming overgrown. Even one season of disuse leaves them impassable because of brambles and growth.

FIGURE 111 Bóthar an Aerstruip, Inis Meáin.

FIGURE 112 Bóthar na gCrag, Árainn. Bóthar na gCrag was built as a Relief Project during the late nineteenth century. It traverses Árainn from east to west and links the crags of each ceathrú. It probably helped access to the most southerly crags, which may have facilitated the individualisation of holdings, but it goes against the north–south organisation of land division. It is somewhat redundant even today, but is a good route for seekers of the cultural landscape who want to avoid the buses and jarveys of the main road.

FIGURE 113 A green road in Eochaill viewed from Bóthar na gCrag, looking north towards the ridge of the island. This road appears on the 1839 Ordnance Survey.

FIGURE 114 A green access road going south from Bóthar na gCrag in Árainn. This does not appear in 1839, but is shown on the 1899 Ordnance Survey. Some of the smaller roads are now becoming impassable through disuse.

FIGURE 115 A road for one cow, Inis Meáin.

FIGURE 116 A typical green road in Eochaill, Árainn.

CHAPTER 5

The Future

Authenticity and Significance

Vulnerability: Changes in Agriculture and Other Threats

The Future: Designation, Stewardship and Management

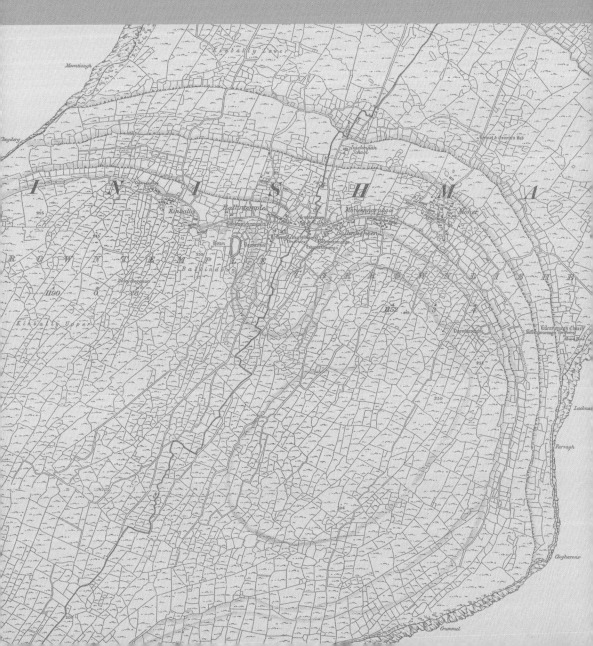

Authenticity and Significance

We have been focusing on the drystone-wall field-boundary system as a separate layer of the cultural landscape in order to tease out its intricacies and understand it more fully. However, in reality it can only be perceived as one layer belonging to the 'deep time' perspective among the other traces of human endeavour and natural elements that make up the cultural landscape. Hence, emerging from this specific study, we look again at the landscape in its wholeness and having considered its history and its making we inevitably find ourselves considering the future of the landscape. Should such landscapes be protected and what does their protection involve? What do they hold for the descendants of the landscape makers and for others? If left unprotected, what is their future? What are the threats posed, do we form policies and how do we manage such a landscape?

The understanding of landscape as a repository of heritage is a relatively recent phenomenon, in particular the idea that an 'ordinary' landscape might be worthy of protection, which has been enshrined in the European Landscape Convention. While the Aran landscape we have been looking at could not be described as ordinary, it has, however, been elementally assembled in an everyday manner by generations of communities. It was and continues to be an agricultural and working landscape.

Before looking at what safeguards may be necessary for the agricultural landscape it is worth looking at what existing protective instruments are in place. The islands of Aran, because of their outstanding natural beauty, are regarded as having a high landscape value, and this is reflected in existing planning legislation. The County Galway Development Plan assesses the Landscape Value Rating as Outstanding, (Draft 2009–2015) Landscape Sensitivity is Class 5: Unique. The islands also have an important archaeological heritage, which has been appreciated and documented and is protected by the Monuments Acts. The geological landscape is widely recognised as having very fine examples of karst, karren and other features characteristic of a limestone terrain, and these features alone would merit particular attention, even if other layers of the cultural heritage were not present. They are recognised as priority habitats. This particular geology combined with agricultural practices has contributed to the abundant and unusual flora that has attracted attention from botanists since the nineteenth century. The islands also have certain species of fauna that are rare elsewhere. The importance of these aspects of the cultural landscape is reflected in the designation of the islands as Special Areas of Conservation (SAC) by the Department of the Environment under the EU Habitats Directive. The waters around Árainn are also part of the SAC. Excluding some pockets along northern coasts in all three islands they are also designated Natural Heritage Areas. The area of the western, eastern and southern coastline of Árainn is a Special Protection Area (SPA). The islands, although recently proposed by Galway County Council to the Department of Environment, Heritage and Local Government, have not been included in the Draft World Heritage Tentative List. At present, however, the agricultural landscape, created incrementally over centuries by the farming and fishing communities who lived on the islands, is not specifically protected nor is there a plan for the future which would secure both the tangible and intangible aspects of this cultural landscape.

When we speak of protection of heritage we are obliged to make an assessment of the value of that heritage: to ask what significance it has and for whom. Is it authentic within the reference terms of its own domain? Does it have integrity in the sense of being intact, or has it been degraded by, for example, unsympathetic development to the extent that it can no longer be perceived as a historical entity? We sense that it has a cultural value, both social and aesthetic, for several reasons related to national heritage, and to the history of both the islands and the country, and we can test that belief in the global arena within the context of existing charters and legislative instruments.

Important guidelines are given in the body of international charters drawn up by ICOMOS (International Council of Monuments and Sites), which are helpful in the assessment of significance, authenticity and integrity, and in the establishment of frameworks for the future protection of heritage. Curiously, the charters which provide most assistance in this case are those dealing with non-European heritage. This is due in part to the fact that the early culture represented in the landscape in Ireland, despite being European, is non-Roman and non-urban. The imprints of the Gaelic world belong to a culture that had become marginalised within Europe at the end of the medieval period and was vanquished before the middle of the seventeenth century. In later centuries the heritage expressed in this particular cultural landscape under study is that of a poor and marginalised community. For these reasons, the charters that we find most appropriate against which to measure the significance of the cultural landscape in question are the Nara (Japan) Document on Authenticity (1994), which sought to broaden the understanding of patrimony, and the Declaration of San Antonio ICOMOS 1996, which dealt with the heritage of the Americas including that of the indigenous people who had a powerful relationship with the landscape. It is a worthwhile exercise to cross-reference some of the important aspects that have emerged from this study of Aran with the elements of these documents.

CULTURAL IDENTITY

The diversity of cultures and heritage in our world is an irreplaceable source of spiritual and intellectual richness for all humankind. The protection and enhancement of cultural and heritage diversity in our world should be actively promoted as an essential aspect of human development. (Nara Document Art. 5)

When taking into account the value of heritage sites related to cultural identity the Americas face the global problem of cultural homogenisation which tends to dilute and erase local values in favour of those that are being advanced universally, often as stereotyped illusions with commercial ends. (San Antonio Declaration Art. 1)

The cultural landscape of Aran is a unique combination of the forces of humankind and nature. One could say that the imprint has been very great there compared to other agricultural landscapes. The inhabitants of the islands have constructed much of the soil cover over the centuries, and enclosed the small fields with walls, so closely together that the landscape is intensively marked by the activities of the community. Yet the built environment of field enclosures and settlement responds so keenly to the natural environment that the boundaries between nature and artefact seem to blur. This kind of response to the natural environment reflects a community living in close contact with natural forces, a relationship with the natural world that is rare in contemporary life.

Today, Irish society is rapidly becoming an urbanised one but, as we have seen, early Irish society had a specifically rural nature and the majority of the population remained in the countryside, generally even after the establishment of towns. Consequently, the cultural landscape of the country expresses this heritage. The spatial organisation of society for the purpose of farming, of cultivation of the land, and of living in close contact with the natural environment is the legacy expressed in the cultural landscape. This investigation of Aran shows that the landscape also expresses the social history of recent centuries. The hardship and difficulties of the nineteenth century are evident. This interaction between humankind and the natural world expressed in the landscape is our patrimony. The value and significance of the cultural landscape is as a repository which allows us to access our cultural identity.

AUTHENTICITY AND HISTORY

An understanding of the history and significance of a site over time are crucial elements in the identification of its authenticity. (San Antonio Declaration Art. 2)

The ancient boundary system which is preserved in the landscape and which exists throughout Ireland, as described in earlier chapters, was the spatial dimension of Gaelic society. As we know, the organisation of society in the Gaelic world was of a rural nature and was not focused on urban centres. Equally sophisticated societies in Europe have left a heritage of towns, if not cities, from which their societies were organised. Frederick Aalen suggests that in Ireland pastoral proclivities may 'have retarded an early emergence of urban life, favouring instead the peripatetic ephemeral urbanism of the fair'.[1] However, the organising device underlying the ephemeral urbanism was the ancient land-division system, the vestiges of which are now called the townland matrix. This system, made up of self-sustaining land-units into which the whole country was divided, permeated the landscape by way of boundaries and named places, and survived through the centuries, forming the basis for the parishes, the Norman manors and even the plantation grants. Even though the nature of the society changed, sometimes very dramatically – after, for example, the Ulster plantation – the boundaries and many of the names remained.

Uniquely, the larger unit of the townland matrix, the *baile*, still exists in Aran, whereas elsewhere in the country the fractional units of *ceathrú*, *cartrún*, etc., have become the modern townlands. The *ceathrúna* boundaries, which subdivide each *baile*, also still exist, generally, and the names of these land-units are known. The *baile* and *ceathrúna* boundaries alone are worthy of conservation, given their relationship to the countrywide scheme and their provenance. For this reason these boundaries have a national significance. The land-division system can be said to be intact and authentic in this location.

MATERIAL AND AESTHETIC VALUE

> *The material fabric of a cultural site can be a principal component of its authenticity. As emphasised in Article 9 of the Venice Charter the presence of ancient and original elements is part of the basic nature of a heritage site.* (San Antonio Declaration Art. 3)

AN ANCIENT CRAFT

The tradition of drystone wall construction is an ancient one and is evident in the still existing great forts from the late Bronze Age and Iron Age and also in early Christian monastic buildings. The field-boundary walls derive from this ancient method and consequently exhibit a continuity with the distant past. There is also evidence that the use of drystone walls as field boundaries dates as far back as the Neolithic period, such as those documented in the Céide Fields of County Mayo. There are prescriptions for drystone wall fences in the early Irish law texts of the seventh and eighth centuries.[2] Belonging to an ancient tradition gives the walls and artefacts a historic and cultural significance. It is an interesting dilemma for the conservator because a drystone wall cannot be repaired in the way that a mortared wall can be: it can only be rebuilt. In order to rebuild, it is necessary to understand the method of construction. In the case of the massive drystone walls of the forts and monastic buildings, aspects of the method are lost to us. At this juncture, the methods of building drystone field-boundary walls are still known, because there are those still living who practise the craft. However, implied in this is the understanding that the craft itself must be learnt and preserved if we wish to conserve the artefacts.

CRAFT VERSUS FINE ART

The aesthetic value of the stone artefacts and drystone walls belongs to the value system of folk crafts. Value systems can be subjective, and it is certainly a subjective reaction to say that one or other wall or stone trough is very beautiful. However, it is impossible to think of a wall or stone trough or gap which is ugly. All of them exhibit the beauty of simple handmade constructions in a natural setting, which are made purely for utility, with no self-consciousness on the part of the maker about making an object of beauty. This is how the craft object differs from a work of fine art whose maker is known, and for whom it is almost impossible to make an unselfconscious piece of work.

MATERIAL

The restricted palette of materials available for use, often self-imposed by fine artists in a search for purity of expression, was a simple reality of life on Aran. There was no other material but stone and this in itself lent ingenuity and beauty to the craft object. A unity of design emerged from the paucity of materials. The *bearnaí* (gaps) are particularly good examples of this. If there had been trees on Aran, gates would have been made of wood and the gap would have never developed. The structural ingenuity of some of the walls, which demanded of the wall builder an intuitive understanding of structure and materials, gives the walls their character. This kind of structural expression is absent from most field and garden walls, which are built according to a structural formula that has been worked out mathematically beforehand. It is a curious paradox that the work of the unknown craftsman can have more individuality and character than that of the fine artist whose work is circumscribed by education and training.

UTILITY

The fact that the artefact made by the craftsman is used by human beings and also, in this case, farm animals, gives the object an intimacy and a patina of use extending over decades or centuries, which adds to its beauty. That it is made simply for use also lends the craft object an integrity and authenticity that demands appreciation.

SETTING

This ordinary landscape of walls, fields, water troughs etc., constructed as part of everyday life and for the requirements of a working farm, has its setting in a landscape of extraordinary natural beauty. The dramatic backdrop of the natural landscape and the heightened quality of light there, reflected as it is from the sea, enhances the visual experience and therefore the appreciation of the viewer. It is impossible to imagine these artefacts without their setting. They have important local significance not just because of the tangible quality of their appearance and use, but also because their intangible value resonates in the community as a result of the many anecdotes and stories associated with them.

SOCIAL SIGNIFICANCE: SETTLEMENT AND LANDHOLDING PATTERNS

> Beyond the material evidence, heritage sites can carry a deep spiritual message that sustains communal life, linking it to the ancestral past. This spiritual meaning is manifested through customs and traditions such as settlement patterns, land use practices, and religious beliefs. (San Antonio Declaration Art. 4)

On Aran, settlement patterns display a relationship between farming community and farmland that is likely to have its roots in the medieval period. The pattern of landholding records traces of farming in common that belonged to earlier traditions and manifests traditional methods of farming, some of which continue today. Despite changes in modes of farming in response to the demands of the time, the overall landscape pattern expresses vestiges of the past. While these patterns, still evident in the landscape, are valuable historical records of past settlement modes and farming traditions, they also engender a deep sense of belonging and allow present-day inhabitants to identify with the past.

This continuity in the landscape, preserved in Aran because of the relative inaccessibility of the islands, and the peripheral nature of the location in relation to the centres of power during the last three and a half centuries, has contributed to the continuation of traditional ways and modes of thought. The survival of the Irish language is undoubtedly another contributing factor which affords a connectedness to the landscape, in particular with named places, something that is often broken when a linguistic shift occurs.

The landholding pattern also shows the gradual individualisation of holdings and the movement away from the practice of farming in common. The intensity of subdivision of the fields in the nineteenth century reflects the hardship of that time and the influence of the landlord system. In this sense, the cultural landscape is a communal monument, which records the struggle of the people to combat adverse conditions. It records their intensity of interaction with the landscape in order to survive the hungry years of the nineteenth century, and their ingenuity in response to the harsh conditions imposed by the political and economic structure of island life. Robinson refers in the introduction to an early map of Aran to the thousand or more miles of walls as 'Aran's most amazing and moving monument put together stone by stone during the centuries of exploitation'.

The rare and significant cultural landscape that we see today holds the collective memory of the period. Equally the landscape tells a story that has resonance in other parts of Ireland; in this sense it has a wider significance.

STEWARDSHIP OF A DYNAMIC SITE

Dynamic cultural sites, such as historic cities and landscapes, may be considered to be the product of many authors over a long period of time whose process of creation often continues today. This constant adaptation to human need can actively contribute to maintaining the continuum among the past, present and future life of our communities. (San Antonio Declaration Art. 5)

In general the island communities who created the cultural landscape still live and work there. This makes the care of the landscape less complex than it would otherwise be. For as long as the practice of traditional farming continues, the islands will remain a dynamic cultural site, even though it is clear from looking back over the history of the islands that the variety of farming practices and the intensity and diversity of farm use are less than they used to be. Some fields and green roads are already falling into disuse. The careful husbandry and hard work of caring for the farm is no longer necessary to provide an adequate income. Servicing the tourist industry puts demands on the farmers' time, particularly during the summer on Árainn and Inis Oírr. However, islanders are aware that if farming were to discontinue altogether, the connection between island and people would be lost.

Dunford points out that each parcel of land in the Burren is managed differently because of its specific ecological, climatic and agricultural qualities, and that, because of the continuity of tradition in these areas, where the same families have farmed the land for hundreds of years, a vast store of management expertise has developed.[3] A continuity of management practices is also present on the Aran Islands: an intimate knowledge of the land exists, such as the particular attributes of each field, the times of the year when it is most advantageous to move animals, and a detailed understanding of what nourishes animals and at what times. All of this knowledge has been passed down from father to son over centuries. The islanders of this generation are fully aware of the myriad twists and turns of places and their associated stories and cultural histories. The challenge is to retain this knowledge and transmit it to future generations. It is part of the intangible cultural value of the site.

AUTHENTICITY AND ECONOMICS: TOURISM, FISHING, FARMING

All cultural landscapes exist in time and, as such, are subject to changing conditions and demands. Although fishing and traditional farming practices are still part of the economy of the islands, tourism is a reality – and at present a necessary part of the livelihood of Aran's smallholders. This is particularly true in Árainn and Inis Oírr, if less so in Inis Meáin, although a new pier has recently been built at the middle island, which will allow and attract more visitors. There is a delicate balance between reaping the benefits of tourism while still maintaining a dynamic authentic landscape. Freezing the landscape and way of life in order to protect the historical and cultural legacy is not an adequate response to the challenge of a vibrant living landscape. Admittedly these are difficult challenges. The answer may be in finding new ways of practising the traditional agriculture that maintains the landscape, supports biodiversity and retains the living intangible heritage of knowledge, garnered from successive generations engaged in the crafts and methods of traditional livelihood.

When we compare the values of this cultural landscape with a global standard of authenticity and significance, we find that it is a cultural landscape of authentic and distinctive character and of significant cultural, aesthetic and ecological quality. For the islanders and others it is a touchstone of cultural identity: an understanding of its evolution gives access to the past, to the ancestral world and to the potential source of spiritual connection. The material fabric inspires admiration and respect. However, the future could see these qualities eroded, in particular if traditional agriculture declines. This is the key challenge, to conservators and primary inheritors alike: to find a way to balance the forces of change with the requirements of heritage, the continuance of a way of life and the natural world that it supports.

Vulnerability: Changes in Agriculture and Threats to the Cultural Landscape

We have been looking at a landscape that continues to be a living dynamic agricultural landscape. Consequently, any consideration of its future must locate it, in the first instance, within the context of Irish and European agricultural practice.

Huge changes have occurred in European agriculture in the last fifty or sixty years. The improvement in efficiency and in methods of food production across Europe has led to a situation where 50 per cent of European farmers now produce 95 per cent of the food; the other half are therefore insignificant in economic terms.[4] However, as Dr John Feehan points out, when writing about the history of farming in Ireland, in conservation terms these are the more significant farms, both for the natural biodiversity supported by their less intensively-managed holdings and for the continuation of traditional farming methods that preserve the agricultural landscape.

Agricultural practice in Ireland began to be affected about forty years ago, particularly after our accession to the European Economic Community in 1972 (now known as the European Union). The Common Agricultural Policy (CAP), which had set out, after the food shortages of the Second World War period, to assure the availability of food to European citizens at reasonable prices, was so successful that it led to overproduction of food accompanied by widespread abuse of the environment. Many of the traditional agricultural landscapes of Europe, including some in Ireland, were lost in the drive for more efficient agriculture during this period of intensification. In Britain, dramatic loss of semi-natural agricultural habitats and species occurred between 1930 and 1984 due to these changes.[5] The realisation that increased agricultural production had begun to work against the long-term sustainability of the countryside was one of the factors that led to reform of the CAP.[6] In 1992, reform measures included policies designed to maintain sustainable agricultural systems, conserve water, soil and genetic resources, and minimise chemical input in order to re-establish the natural equilibrium of soils and plants, and to promote diversity in fauna and flora.[7] Further reform in 2005 ushered in a new agenda, which ended support payments for agricultural production but emphasised the keeping of land in good environmental condition.

FIGURE 117 Cattle resting on the limestone pavements of Cill Éinne, Árainn.

Aran was saved from some of the abuses that resulted from the CAP regime during the 1970s and 1980s because of its isolation, the difficulty of using machinery there and the particular nature of the landscape and farming practices. However, the damaging effects of fertiliser use in a karst landscape quickly became obvious in those years as water sources on the islands became polluted. This was also a serious problem in the similar geological landscape of the Burren where more than 50 per cent of springs and wells were contaminated.[8] Parts of Árainn were overgrazed and damaged by agricultural improvement.[9] Aran farmers benefited from the system of subsidy payments that has now been dismantled under the CAP reform, but they will continue to benefit from the Single Payment System so long as a minimum number of stock is kept on the land.[10] This feature undoubtedly allows smallholders to continue farming on land that would otherwise have fallen into disuse or have been merged with larger farms. The reduction in numbers of people working on the land and the increase in size of holdings evident throughout the EU was mirrored in Aran. The Department of Agriculture estimates that the number of farms on the islands has decreased by more than 30 per cent in the last fifteen years. Likewise, Aran has followed the pattern of the rest of Ireland with a decrease in diversity of livestock, the main farming pattern being dry stock with some suckler cows. There is no dairying and very few sheep; some farmers have Connemara horses on the land, and there are some goats. There are 214 registered herd owners on the three islands, most of whom are subsistence farmers with herds of fewer than ten animals.[11] They supplement their farming income with fishing and tourism activities.

The primary threat to any agricultural landscape is that of disuse; a secondary threat is 'improvement' or intensification. While reduction of use or 'under-management' is the most significant threat facing this landscape at the present time, 'improvement' could also be a threat, in particular to the stone artefacts. Both situations would affect the rare flora and fauna of the islands. In the past it was commonly believed that to leave a landscape without human interference of any kind was the best way to protect the flora and fauna. Conservators have now realised that there are often strong links between wild biodiversity and traditional agriculture. In Aran the protection provided by the stone walls, combined with the unusual transhumance practice of wintering cattle on the crags, creates an environment in which these unusual flora can flourish. The small fields and traditional farming practices are recognised by the National Parks and Wildlife Service as vital to the preservation of this Special Area of Conservation.[12]

THE RURAL ENVIRONMENT PROTECTION SCHEME

In terms of conservation, the most important measure in Ireland to result from the CAP reform is the Rural Environment Protection Scheme (REPS), which came into operation in May 1994. In July 2009 entry to the most recent scheme, REPS 4, was closed by the Department of Agriculture. REPS recognises the pivotal position that the farmer holds relative to food production and interaction with the environment, and aims to encourage farmers to continue their age-old role as guardians of the countryside as well as producers of food.[13] The primary aim of the REPS is to 'provide an aid scheme for farmers to encourage the introduction of farming practices and production methods which reflect the increasing concern for the conservation of wildlife habitats and endangered species of flora and fauna and the protection of the landscape'.[14]

The environmental objectives also include the production of food in an environmentally responsible manner, the protection of the agricultural landscape and the safeguarding of the resources of the countryside such as watercourses and soil equilibrium. The scheme also encourages breeding of endangered species of livestock animals.[15] REPS is administered by the Department of Agriculture Food and Forestry and is a voluntary participation scheme but, as we shall see, there is high level of participation in Aran.

The REPS has a particular role in areas such as Aran, where conservation of the natural environment and of the heritage of traditional farming has outstripped the importance of agriculture. There are at present 200 viable REPS applications from the Aran Islands[16] – in other words, almost all of the farmers are involved with REPS. This widespread participation in the scheme is due mainly to two factors. Firstly, the islands, under earlier legislation, were Natural Heritage Areas (NHAs). They are now Special Areas of Conservation (SACs) under the EU Habitats Directive administered by the Department of the Environment. Farmers whose land occurs in such areas are obliged to join the REPS scheme or consult with the Department of the Environment on how they will comply with the requirements of the SAC designation. They are entitled to financial compensation for loss of income due to altering farming methods under the Habitats Regulations.[17] In practice, they are offered an opportunity to join the REPS and additional payments are made because of the SAC designation.[18] This makes participation more financially rewarding in a SAC than elsewhere. Aran farmers generally have very small holdings and they need the financial support to continue farming. Secondly, farming on the islands has, in general, remained traditional and has not engaged in the intensive farming methods pursued by earlier CAP policies, with the result that Aran farmers do not have to alter their methods very much in order to participate in the scheme.

When farmers join the REPS, the first step is to produce an 'agri-environmental plan'. This is a management plan for the farm that is agreed with the REPS planner, generally an agricultural scientist or farm planner with little previous experience of conservation. The absence of qualified ecologists from the approval process has been criticised.[19] Although an environmentalist or ecologist was required to advise in the case of agri-environmental plans being drawn up in SACs or NHAs, under earlier REPS,[20] this is no longer the case. The plan deals with eleven different areas, which include the maintenance and repair of farm and field boundaries and protection of features of historical and archaeological interest.[21] In addition, if the land is within an NHA or SAC, the farmer must follow management guidelines outlined by the National Parks and Wildlife Service. These include measures to do with scrub clearance, stocking and fertiliser use. The farmer undertakes to follow this management plan and s/he is paid an amount per hectare per year for a five-year period. There is an upper limit of 40 hectares (98.84 acres) for the full payment; in excess of that amount a supplementary payment per hectare is made, but the REPS measures must be carried out on all land if the farm exceeds 40 hectares.[22] In the event of the management guidelines not being followed, a financial penalty is incurred. While the plans are drawn up with the advice of Teagasc or an independent adviser, monitoring is carried out by the Department of Agriculture. Monitoring of compliance with the SAC is carried out by the National Parks and Wildlife Service.

The REPS farmer also makes a commitment to twenty hours of training/education per year. This is an important aspect of the scheme. In the islands, this takes the form of a seminar which takes place over a few days each year at which an ecologist/environmentalist, an archaeologist/conservationist and a Teagasc or REPS adviser give talks and have a dialogue with farmers on the requirements of the scheme.

ASSESSMENT OF THE REPS AS A CONSERVATION INSTRUMENT

REPS, combined with legislative protection under the Special Area of Conservation/Habitats Directive, provides a modicum of protection at present in the absence of other measures, not least because it allows farmers on smallholdings to continue to use traditional methods in an environment of economic viability. Although it does not set out to conserve it for historical or aesthetic reasons, in fact it protects the drystone-wall field-boundary system by default, and holds at bay the primary threats to the agricultural landscape – disuse and intensification or overuse.

It is, however, a stopgap measure as an instrument of conservation. Fortunately, it has been in place during a time when many walls could otherwise have been removed, but the closing of entry to REPS 4 confirms that there is no guarantee that it will continue into the future. In addition, the protection of the stone walls and artefacts is not actually written into the scheme except as part of the maintenance of field boundaries. There is no specific REPS policy written for the islands, despite the unique conditions there. An additional list of prescriptions to the basic REPS has been agreed for the Burren, though it does not include references to historic built fabric on the farms.[23] In theory, the REPS is a blueprint written in general terms for the whole country. In practice, it is tailored to suit the local environment and the needs of particular areas.

However, the scheme could continue to function well if it was part of an overall conservation and management policy for the islands. It would have to be supported by research and documentation and would benefit from being connected to a global conservation network. At present, much depends on the sensitivity, understanding and goodwill of the REPS adviser. The attributes that make it an interesting and successful implement for conservation of the cultural landscape are: firstly, it goes to the heart of the problem by making the continuance of traditional farming a viable possibility for farmers; secondly, it has a local reach, and therefore has the ability to change to suit the local situation; thirdly, it very directly involves the primary inheritors – the islanders.

The scheme needs to be supported by an advice document that refers to the specific historic fabric and the means of safeguarding the significant elements of the cultural landscape. A specific plan would include issues such as: ancient boundaries, stone artefacts; *céimeanna* and *staighrí*; stone water troughs; abandonment and scrub; a record; and education.

ANCIENT BOUNDARIES, STONE ARTEFACTS

At present, the REPS plays a role in the conservation and maintenance of stone boundary walls, as part of its overall aim to 'keep farms and farmyards tidy' and to contribute to the scenic character of the landscape.[24] However, there is no differentiation between walls, or reference to the heritage value or historical and aesthetic importance of certain walls, such as the *ceathrú* and townland boundary walls.

Similarly, there is no particular protection for the artefacts associated with the walls, such as the *bearnaí* (gaps). In fact, there was a requirement in REPS 1 to provide a gate in every field. This was not enforced in Aran because the advisers realised how insensitive that would be to the traditional Aran field.[25] Yet some gaps are being lost in favour of gates. Although regrettable, it is understandable why many farmers now replace the *bearnaí* with gates. One can see how, for the working farmer, it is much simpler to use a gate than to have to build up the gap with stones every time s/he moves livestock. REPS advisers now guide farmers towards simple wooden gates if the gap is being replaced.[26] The guidelines should be extended to require that gaps be kept even if a gate is required for ease of movement around the farm.

CÉIMEANNA AND STAIGHRÍ

The two types of stiles described in Chapter 4 are not at present under threat because they still perform a function that is not fulfilled by the gate. However, there should be a reference to stiles which would stress the importance of their conservation. Otherwise, they could easily be lost in a short space of time during a spate of energetic maintenance of walls.

STONE WATER TROUGHS

REPS give clear guidelines for farming activities in the vicinity of archaeological monuments. These guidelines could be extended to include the conservation of field furniture, so that all remnants of the agricultural landscape heritage would be conserved. Water troughs in recent years have sometimes been made of concrete instead of stone. (At one time, the Department of Agriculture grant-aided concrete tanks.) Some Aran farmers certainly recognise the importance of the stone water storage troughs as cultural artefacts,[27] and view the ingenuity and craft shown in the making of these as a tribute to past generations, a memorial to the ancestors of today's Aran farmers.

FIGURE 118 Bloody crane's-bill (*Geranium sanguineum*) on the crags, Árainn.

ABANDONMENT AND SCRUB

The control of scrub is a complex problem that has to be addressed and resolved if the species-rich semi-natural grasslands that evolved through traditional farming practices are to remain.[28] The presence of scrub also has a detrimental effect on birdlife. Dunford points out that the magnificent flora on the Burren uplands is unlikely to survive without the presence of wintering cattle.[29] Equally, if cattle were to be sent there in large numbers for pasture in the summer, which would now be possible due to supplies of piped water, the flowers would be eaten before flowering or seeding. The same must be true for the flora on the limestone pavements of the upland regions of Aran, such as Na Craga, which were traditionally and still are used for wintering cattle. Likewise, the small walled access roads, some of which were constructed before the 1839 Ordnance Survey, revert completely to scrub if not used. Some are already impassable.

This problem has been identified as a major issue in recent studies of the Burren.[30] Changes in the breeds of cattle, wintering younger cattle on the uplands and, in the view of some Burren farmers, the short space of time allowed for wintering by the REPS are some of the factors regarded as contributing to scrub invasion. Once scrub has taken hold, the fields are no longer suitable for cattle. The keeping of goats may be an answer but would require a market for kid meat and cheese. The keeping of older cattle (which are better able to graze on tough branches) would also contribute to scrub control but, again, a market would be required for such animals. At present an agri-environmental pilot scheme is being successfully run in part of the Burren with EU and Irish government support. This scheme has introduced cows and cattle to conservation areas where they graze at specific times to control scrub and preserve the wild flower species for which the Burren is famous.

A RECORD

At present, the record that exists is the Ordnance Survey aerial photographs of 2000, which shows location of walls and artefacts. The serious problem of scrub encroachment cannot be solved until a clear record is made of the rate and extent of scrub. In the absence of a record, monitoring of the REPS by the Department of Agriculture is problematic.

EDUCATION

The REPS training programme is a good one and the importance of this aspect of the scheme should be stressed. There is also the question of educating the next generation in the farm-management practices of the island. Knowledge of the landscape garnered over centuries, management of the land, animal husbandry and the crafts of drystone wall building, stile building and land-making will otherwise be lost. The teaching of such skills could be incorporated into the curriculum of the Colaistí Gaeilge which host large numbers of young people every summer. Teaching of skills such as the removal of scrub by hand, wall building and stone trough-making could be part of the Irish language and craft that young people learn on the islands. The National Museum already runs similar educational programmes in the Museum of Country Life in County Mayo during the summer. Workshops in stone wall building have also been run by Galway County Council. Such programmes would be a way of maintaining and caring for the cultural landscape as well as passing on some of the skills and knowledge to the younger generations. As the poet and philosopher John O'Donohue, native of County Clare, says, referring to the Burren: 'There is a world in the land, a farming world of the most sophisticated complexity and the most astute and rich memory that in the next ten years will have vanished completely. Isn't there something wrong with either our way of life or style of education that these huge ventricles of life, of memory or of perception are not being passed on?'[31]

Fuller and more open discussion with the islanders is also necessary for any workable conservation strategy. Some regard REPS as 'a scheme from the top down' and some farmers feel they should be more involved in a consultation process. Clearly, the primary stakeholders are the islanders themselves and this landscape is a memorial to their ancestors and a landscape of their own making. Any move forward into a strategy for the future must be made with them.

To some extent there appears to be something of a blind spot about threats to the landscape in Ireland. A recent report of Landscape Character Assessments (LCAs) found that of the twenty-eight LCAs completed at the time of the report (2006) only seven identified threats to the landscape. Of these seven, none identified road building or infrastructural projects as a threat. Admittedly some of these LCAs were written five or six years earlier, at a time when road building was not as evident as it was later to become, and clearly it is difficult to identify threats before they appear; however, that unfortunately is precisely what has to be done. The unforeseen threat becomes the most potent when it arrives.

Lack of knowledge among decision makers also presents a threat to the cultural landscape. The current County Galway Development Plan recognises Kilronan as a 'settlement area' but does not recognise the other villages as such. Yet permission is given for houses to be built at a distance from the villages, which is antithetical to the traditional settlement pattern. This indicates a lack of understanding of historical and heritage issues. Drystone walls bordering the road are routinely demolished as a result of a condition attached to the planning permission to provide parking space in front of a new house. There is also generally a condition requiring that walls be rebuilt; however, historic fabric is being lost every time this happens. Each small loss of historic fabric and each occurrence of unsympathetic or antithetical development adds to the overall degradation of the landscape and eventually the cultural landscape loses its integrity.

The Future: Designation, Stewardship and Management

Notwithstanding that the islands have been excluded from the 2009 Draft Tentative List of World Heritage Sites, the cultural landscape of the drystone-wall field-boundary system may have met the criteria of subcategory (ii) – organically evolved landscape – World Heritage Cultural Landscape. The importance, however, of focusing attention on the possibility of inscription as a World Heritage Site is the recognition that an overall management plan for the future of the islands is required.

There is a degree of urgency about putting such a plan in place for the landscape, whether or not inscription to the World Heritage List is being considered, because of the precarious balance between traditional agricultural use and the development of tourism, the latter being a factor contributing to abandonment of the land. Despite the current measures which exist within REPS, each year the amount of agricultural work done lessens in favour of servicing the tourism industry; little by little, scrub increases on the small access roads and even in some of the fields; livestock kept on the farms becomes less diverse; fowl are generally no longer kept. The invasion of scrub will have a negative effect on birdlife, and without cattle wintering on the limestone plateau the rare and interesting flora that exist there will diminish, if not disappear. The overall effect of these incremental changes is that biodiversity and cultural diversity are gradually eroded with a homogenising effect on landscape, culture and nature: the rich and diverse world of the cultural landscape begins to fade.

There are, in different parts of the world, cultural landscapes which share parallels with Aran. The extraordinary Rice Terraces of the Philippine Cordilleras though far greater in extent than the Aran field system have some similarities: they represent a communal monument created by hand in a remote and isolated location for the purpose of agriculture – rice growing in this case; they are a unique and remarkable example of the harmonious interaction between humankind and nature created over many generations under harsh conditions; a rich oral tradition informs the methods of construction of the terraces, care of the paddy fields, irrigation and farming methods, and remarkable seed variety (more than 500 varieties of rice suitable for high altitudes), including in this case a ritual heritage of religious significance. Parts of the site were inscribed as a World Heritage Site in 1995. Although the new status brought international attention to the area, problems to do with management, funding, and abandonment and deterioration of some of the terraces led to the site being inscribed on the World Heritage in Danger List in 2001. Some feel that the preparation period for World Heritage listing may have happened too fast.[32] Whether or not this is the case, the reality is that continuing cultural landscapes present difficult challenges, not least because of economic and social pressure on the communities who inhabit them. The terraces, which are particularly important for the agro-biodiversity they support as well as their cultural value, are now being managed under a pilot scheme (2007–2014) 'to help national and local stakeholders protect and sustainably conserve the systems and their components' which is part of the Globally Important Agricultural Heritage Systems (GIAHS) run by the Food and Agriculture Organisation of the United Nations.[33]

Another site with some similarity to Aran is the landscape of the Pico Island Vineyard Culture of the Azores, in the Atlantic Ocean. Along the northern and western edges of this volcanic island is a field system of high drystone walls of volcanic rock enclosing tiny soil-less fields (some a few metres square) for the purpose of viniculture. The walls protect the crops from Atlantic winds and salt spray and also provide support for the tendrils of the vines and the fruit. Part of the site is still in use for wine growing, producing a highly regarded sweet wine called 'Verdelho'. Other parts are a relict landscape which is overgrown but could be renewed. The farming system may date back to the fifteenth or sixteenth century and is likely to have reached its peak during the nineteenth century. The small plots are thought to have become extensive during the nineteenth century when social change encouraged numerous small farmers. This site was inscribed on the World Heritage list as a Cultural Landscape in 2004.[34]

Cinque Terre is a coastal area of northwestern Italy encompassing small towns with an associated agricultural landscape on steeply contoured coastal hills. The site has been terraced, forming flat ledges for the growing of vines, which are held in place by drystone retaining walls. The local population have been engaged in this particular form of land use for more than a thousand years. The terracing method controls the flow of water and prevents erosion of the hillside, which is important as the towns are below the area of the farmed land and would otherwise be subject to landslides. The terracing creates a scenic landscape in a setting of great natural beauty, which makes it an attractive destination for tourists. This site was listed as a World Heritage Cultural Landscape in 1997. It is also a Category V Protected Landscape (IUCN). A report carried out in 2006 showed that the site had retained the World Heritage values, and that the traditional lifestyle necessary to maintain an authentic continuing cultural landscape had been maintained despite external pressures from socio-economic forces.[35] A training workshop was held there in 2007 – with the topic Managing World Heritage Cultural Landscapes – to discuss the challenges and experiences of managing cultural landscapes in the southeastern European states including Italy.

All of these sites share similar challenges to those facing the Aran islanders. Even in Cinque Terre, which has been enjoying more successful management than the Philippine example, young people are reluctant to work in traditional viniculture, and knowledge of the craft of drystone retaining-wall building, crucial to the hydrogeology of the hillsides as well as being of cultural importance, is retained by only a handful of older men.[36]

World Heritage listing has raised awareness in all cases of the existence of these special places. The high status of the listing helps to allay certain kinds of obvious threats such as antithetical infrastructural development. The more insidious threats such as the loss of the traditional lifestyle are more difficult to tackle given that they pervade the community for various socio-economic reasons usually through influence from the greater world. The listing also increases tourism and this can be a double-edged sword, when the influx of people is too great and is detrimental to the landscape. Raised awareness within the community also occurs and connection to a global network of equivalent places experiencing similar challenges must be one of the most beneficial aspects. In some places people feel that World Heritage listing is more to do with the conservators (from outside the community) than with them. This is unfortunate because in all conservation situations but particularly in organically evolving cultural landscapes, conservation must come from within; the continuing cultural landscape does not exist without community.

Inscription as Category V Protected Landscapes/Seascapes may be a possible way forward for the protection, guidance and management of Aran as a cultural landscape, if legislation in Ireland allowed for such designation; at present it does not. The IUCN Protected Landscapes is a parallel system to the UNESCO World Heritage List and some sites are inscribed as both. But the most relevant difference in the discussion about the protection of Irish cultural landscapes, particularly those that would come under subcategory (ii) – organically evolved landscape – is that the Category V Protected Landscapes does not seek qualities of outstanding universal value, and therefore cultural landscapes of national importance could be considered for inscription. This is not to say that it is a second-rate protection, simply that it has a different emphasis. This may be more important to Irish cultural landscapes other than Aran, because the islands have qualities that are likely to meet the criteria required for 'outstanding universal value'. There are many Irish cultural landscapes that would not meet those criteria but nevertheless, hold important national or local values. However, much of the work carried out by IUCN conservators in relation to Category V Protected Landscapes/Seascapes seems to have a particular resonance with the Aran Islands. The description of a Category V landscape is as follows:

> protected area managed mainly for landscape/seascape conservation and recreation – area of land, with coast and sea as appropriate, where the interaction of people and nature over time has produced an area of distinct character with significant aesthetic, ecological and/or cultural value, and often with high biological diversity. Safeguarding the integrity of this traditional interaction is vital to the protection, maintenance and evolution of such an area.

Looking at the Selection Criteria for Category V Protected Landscapes vis-à-vis the Aran landscape shows that the landscape can meet the criteria very well. The following is based on a similar exercise relating to the Philippine rice terraces of the Ifugao people:[37]

SELECTION CRITERIA FOR CATEGORY V PROTECTED AREAS (BASED ON PHILLIPS *ET AL*)

ESSENTIAL CHARACTERISTICS	CHARACTERISTICS OF THE DRY-STONE WALL FIELD BOUNDARY SYSTEM OF THE THE ARAN ISLANDS
Landscape and/or coastal and island seascape of high/ distinct scenic quality	The field system created by the islands' inhabitants over hundreds of years presents an intricate and fascinating land mosaic in a setting of extraordinary natural beauty, situated as it is on islands off the west coast of Ireland in the Atlantic Ocean. The shallow soil layer of fields created by human beings is spread thinly over a remarkable limestone geology which rises from the shallow northeast coast to the southwest (about 120 m at the highest) presenting spectacular cliffs to the ocean.
Significant associated habitats, and flora and fauna	Rare species of plants/wild flowers/grasses occur on the limestone pavement, due partly to the unusual form of transhumance practised there among farmers, also to the shelter afforded by the closely knit stone-wall boundary system. Rare coastal species also occur at the cliffs, dunes and submerged areas. The birdlife is considered of international significance due to the numbers of species present which are rare in Europe. The Common Seal and a rare snail (*vertigo angustior*), both threatened species, are also present.
Evidence that a harmonious interaction between people and nature has endured over time and still has integrity	The humanised landscape, constructed over the last few centuries for the purpose of agriculture with roots in the more distant past, expresses the close relationship between community and the natural world. This landscape is still used by the farming and fishing community although income is supplemented by other activities, mainly in tourism.
Unique or traditional land use patterns, e.g, as evidenced in human settlements	The islands display a good example of the ancient land-division system in Ireland. The relationship between sustainable land-unit and village, and between fractional land-unit (*ceathrú*) and township (*baile*) belongs to the Celtic system of land use. The system, though existing on the mainland of Ireland, appears more intact on Aran. The traditional seasonal pattern of land use is still practised by farmers. The unusual form of transhumance where cattle are moved to the higher areas of the limestone pavement during the mild oceanic winter is also practised in the similar limestone geology of the Burren in County Clare.
Valued for the provision of environmental services, e.g. watershed protection	Water collection at the inlands cliffs in natural and man-made wells was the traditional method of freshwater provision for human consumption. These still exist and their location underlies the pattern of settlement. Fresh water is now stored in large central reservoirs on each island and piped to the villages. Traditional rainwater collection tanks made from stone are still used for watering animals.
Valued for sustainable use of natural resources	The making of land using the traditional method of gathering and mixing seaweed and sand, and the protection of soil cover by the sheltering stone walls has created and sustained much of the fertile land of Aran.

Unique and traditional social culture as evidenced by local customs, livelihoods and beliefs	The most important cultural aspect is the continuing use of the Irish language as a first language among the majority of islanders. Consequently, place names, names of fields, farm implements and methods are preserved in the language. Many local customs associated with farming and with the sea still exist, although they are at risk of being lost. Crafts associated with the farming livelihood and stone wall building, for example, though still extant, could be lost in the next generation. Making their own knitted and woven clothes is a craft that is almost lost, although some older men still wear the traditional outfit. This craft has inspired contemporary knitwear businesses.
Opportunities for public enjoyment through recreation and tourism consistent with lifestyle and economic activities	Tourism is already quite a developed activity on the islands; people are attracted to the archaeological sites as well as the natural beauty. Most islanders derive some economic benefit from it, more particularly on Árainn and Inis Oírr, although the recent construction of a second pier at Inis Meáin may see more tourists to that island. Irish language summer colleges for school students from the mainland also exist.
Suitability for scientific research	The islands are of considerable scientific interest with ample areas for research in their array of good quality habitats supporting many rare and threatened species, diversity of limestone pavements and other karstic phenomena, rich archaeological heritage and traditional farming ecology.
Important for education	The role of the islanders as educators of the Irish language has been recognised since the era of the Gaelic Revival at the end of the nineteenth century, but there are also other aspects that make the islands important for education. Among these is the ingenuity displayed particularly during times of rising population and limited land resources. Common to other indigenous or traditional communities much can be learned about sustainable land use in Aran.
Recognition of artists of all kinds and cultural traditions (new and from the past)	Many artists in the past found inspiration on the islands both in the natural and wild beauty of the setting and in the livelihood and traditions of the people. The island communities have also produced widely recognised artists of their own. The tradition of the islands as a source of inspiration for writers, painters, photographers and others continues today.
Important for agro-biodiversity	A level of agro-biodiversity exists consistent with present-day traditional farming methods on small farms, but much less than formerly.
Potential for ecological and landscape restoration	Potential exists to renew agro-biodiversity to former levels. A small number of organic farms exist which have returned to a diverse livestock rearing and vegetable growing, but more are needed to maintain a delicate balance. The key challenge is the continuance and sustaining of the traditional agricultural practice. It is intrinsically linked to the conservation of wildlife, biodiversity, heritage and the cultural landscape itself.

The first eight criteria are considered essential, the remaining five are considered desirable. The weakest link may be the 'importance for agro-biodiversity', a desirable characteristic. A high level of diversity in traditional agriculture that supports biodiversity existed in Aran up until at least the middle of the twentieth century, but has been on the decrease since that time. However, there is potential for farm diversity to increase and this would be an important part of any conservation management plan.

The Category V Protected Landscape propose a strong input from the local community: ideally, the impetus to have the site inscribed should come from within the community. Their engagement with the whole process is crucial to the success of the management of the project. Aran is fortunate in that it already enjoys strong community engagement in policy and decision making for the islands. Each of the three islands has a Development Co-operative, which has a voluntary committee and is administered by a development officer. The co-op deals with many aspects of island life from infrastructure to social and economic issues, including, for example, a recently developed ecological waste management and recycling system for each of the three islands. In general, the island people control things to do with the islands, and fight to maintain control over transportation, the tourism industry, the Irish language schools and so on. The islanders generally are very aware of their role as custodians of the Irish language and associated traditions. Their involvement with Irish language education has developed abilities in areas to do with management of their resources. In recent years they have initiated some projects to do with recording aspects of the landscape. In Inis Meáin a local group collected and recorded cartographically all the names of the fields of the island. In Árainn a group of women, Mná Fiontracha, produced a series of pamphlets to do with the oral tradition. One of these recorded the names and locations of wells, roads and paths, and *carcaireacha* – a slope or inland cliff characteristic of the islands – and also maps of the villages with the names of the families who had lived and those who were still living in each of the houses.

There is ample evidence that a management strategy for the islands could be developed from within the community, particularly if people were to realise that serious threats to the integrity of the island landscape exist, the least obvious but perhaps most potent being the slow but sure abandonment of the traditional farming of the land and the creeping homogeneity of the farming type that remains. Creative initiatives are needed to encourage independence and viability in traditional agriculture.

There are lessons to be learned from the nearby cultural landscape of the Burren, where a pilot scheme, BurrenLife, has emerged which puts the farmer at the centre of the conservation process. This initiative offers the possibility that a Burren brand for meat and other food products might be produced in the context of conservation management. It may also be an answer to problems of scrub control and land use.[38] This would seem a possibility well worth investigating in Aran, where the drawbacks of trying to sell cattle from the islands, the costs of transportation and so on, are constantly lamented by the Aran farmers. It is important to remember that most of Aran's farmers have very small stock numbers, fewer than ten head of cattle per farm. In this context, all the farmers of the island would have to group together in order to pursue a viable product. This may signal a return to a form of co-operation between farmers and conservators that would enable new initiatives to emerge, or a new kind of farming in common.

Similar conservation initiatives exist in some of England's Protected Landscapes: national parks or Areas of Outstanding Natural Beauty (AONBs) in which farming is still the predominant land use even though tourism is the more economically beneficial activity. More than a quarter of a million people live in these landscapes. In the Lake District National Park the commonage of the high fells is grazed by the traditional sheep of the area: the Herdwick – a hardy breed that is able to survive on the scant herbage of the uplands. Recent changes in the CAP have had the effect of undermining upland farming with the result that the National Trust has had to seek new markets for Herdwick wool. The meat of the Herdwick is much prized by gourmets because of its gamey taste, which is due to their frugal diet and slow growth. The Herdwick Breeders Association, Friends of the Lake District, the National Park Authority and the National Trust have come together to market the meat.[39]

On the edge of the Lake District National Park in the Lyth Valley a group of local people have come together to form the Westmoreland Damson Association and with EU support they are preserving damson orchards. Before the Second World War damson trees (a type of plum) grew in profusion in the area and the numbers now are greatly reduced. This association is involved in creating small businesses for damson products such as jam, ice cream, beer and gin. They also, with the help of Natural England and the Environmental Stewardship scheme, are re-planting the characteristic linear and field corner damson orchards of the district.[40] Until the 1930s the plenitude of orchards was a characteristic aspect of the English cultural landscape. Virtually every farm made cider from the 1700s onwards, but all of this changed with the industrialisation of agriculture in the mid-twentieth century. Grazing of animals including pigs, ducks and geese occurred under the large apple trees creating a rich environment for biodiversity and agri-diversity. A conservation initiative currently underway in the Tamar Valley AONB in Cornwall in association with the National Trust is the creation of a 'Mother Orchard' which will preserve apple varieties specific to the area.[41]

Other places such as the Shropshire Hills and the South Devon AOBN have been inspired to conserve apple varieties and local communities have grouped together to establish markets for apple products.[42]

What is abundantly clear is that a management strategy, informed by the diverse influences of the past, with a creative approach to the future, and aware of similar cultural landscapes worldwide, is required to safeguard the continuing cultural landscape into the future. While recognising that many aspects of that strategy already exist in disparate form – such as: REPS; educational programmes run by local government and the Heritage Trust; and local initiatives toward recording heritage in the landscape – there is a necessity to raise awareness locally and among external decision makers that threats are at hand. The most potent threat is probably the least evident in that it comes in the form of incremental change in the landscape due to abandonment of the land or insensitive development. Existing initiatives together with new and creative ways of thinking about the landscape should be brought together in an overall strategy. The philosophy at the core of the strategy should be that this layer of the cultural landscape would continue to remain at the heart of the livelihood of the islands, while still providing a place that visitors can enjoy, learn from and appreciate. This philosophy recognises that: the connection between land and people is integral to the conservation of the cultural heritage; the connection is now threatened because of changes in agricultural policies and global economics, among other factors; and that while tourism is an important and welcome part of contemporary island life, balancing the influences and demands of tourism with the traditional lifestyle is the key to maintaining a dynamic continuing cultural landscape.

The importance of the cultural landscape must be recognised in law as an expression of the national heritage and the foundation of the cultural identity of the people, as proposed by Article 5 of the European Landscape Convention (Council of Europe) ratified in March 2004. Local government would recognise the significance of the cultural landscape and incorporate measures to conserve and protect it into regional and local policy, particularly planning legislation.

Thorough and informed multidisciplinary research would also be a part of the strategy, including archaeological investigation, to investigate fully and record the drystone-wall field-boundary systems and the associated stone artefacts. The research would be documented and presented in such a way that it could be used for awareness-raising and education. Training and education would take place on a number of levels so that a true understanding of the cultural landscape would become widespread. Documentation would be made available to decision makers such as:

- planners in Galway County Council, so that they could be fully informed about matters such as traditional settlement patterns or the importance of the stone wall boundaries as part of the cultural heritage;

- officials from the Department of Agriculture who currently monitor the REPS scheme, but who are at present working without a comprehensive record of what exists in the cultural landscape;

- Teagasc and other private companies which prepare the agri-environmental plans with the farmers but do not have access to information sources that describe the heritage and cultural significance of the agricultural landscape.

The curricula of existing educational programmes such as the REPS training seminars and the Irish language colleges would be extended to include study of the cultural landscape. A policy document would be written and monitored over time to address issues specific to the landscape of the islands. Existing workshops offered by local government and the Heritage Council would be encouraged and augmented to initiate the next generation (of young islanders and others) in the traditional crafts of stone wall building, animal husbandry and farm management practices to enable people to identify significance in the cultural landscape, and recognise the forces that threaten it.

Conclusion

I n conclusion it can be said that the drystone-wall field-boundary system of the Aran Islands is one of the many diverse and rich cultural landscapes of Ireland. The islands retain the ancient land-division system, which was the spatial dimension of Gaelic society, and it is remarkably intact there. The settlement pattern and agricultural methods suggest continuity with the late medieval period or even earlier. The intense subdivision of land that characterised the response of the islanders to the difficulties of the nineteenth century is relatively unchanged since it reached its climax at the end of that century.

The implication of this within Ireland is that it is necessary to recognise fully that much of our heritage is expressed in the landscape, and that loss of such heritage erodes our sense of cultural identity. In the context of Europe, the protection and appreciation of such landscapes has relevance because they reflect a world that has been lost in many other places as a result of progress and development in recent centuries. In a broader global context, there are parallels with non-western cultures and with marginalised communities within Europe where societies lived in close relationship to the natural world until they were eclipsed during the colonial era or more recently by powerful global economic forces. In this sense, it engages with the broader debate about cultural patrimony that is taking place within the discipline of conservation.

There is a rich legacy of heritage, history and social history to be gleaned from choosing the landscape itself as a primary source for study. Ultimately, the interdisciplinary approach has been rewarding when all the threads can be seen coming together in the landscape. However, it is clear that a true record of the landscape requires further multidisciplinary work, which would include archaeological investigation.

Awareness of the unilateral changes taking place in agriculture points to the precariousness of the cultural landscapes of Ireland. Those agricultural landscapes not occurring within outstanding natural landscapes are particularly vulnerable. Legislation and education must be developed if we hope to counteract the penetrative forces of change which threaten heritage and cultural identity.

In the final analysis it must be admitted that without traditional agriculture and a community to support it, the drystone-wall field system will disappear – that is to say, the fields will become overgrown and the access paths impassable. In time it may become another kind of wilderness. Alternatively it could become a museum landscape – a static site, tended and cared for but devoid of a living and connected community – a sad end for a vibrant and dynamic cultural landscape. The task is not simply to conserve the landscape but to find new ways of living in and caring for the land, not only learning from the traditional and sustainable practices of the past but also embracing the knowledge gleaned from the now accessible wider world experiencing similar dilemmas. Undoubtedly these are considerable challenges. However, the positivity and energy that exists within the island communities and the high regard and affection in which the islands are held in the rest of Ireland could translate into a new way of moving forward into the future.

References

CHAPTER 1

1   Aalen, F. H. A., 'The Irish rural landscape: synthesis of habitat and history', *Atlas of the Irish Rural Landscape*, edited by F. H. A. Aalen, Kevin Whelan, Matthew Stout (Cork University Press, Cork, Ireland, 1997), p. 5.

2   Aalen, F. H. A., *op. cit.*, p. 14.

3   Aalen, F. H. A., *Man and the Landscape in Ireland* (Academic Press, 1978), p. 80.

4   Aalen, F. H. A., 'The Irish rural landscape: synthesis of habitat and history', *Atlas of the Irish Rural Landscape*, eds. F. H. A. Aalen, Kevin Whelan, Matthew Stout (Cork University Press, Cork, Ireland, 1997), p. 19.

5   Martin, J. & Farmer, A., *Landscape Character Assessment (LCA) in Ireland: Baseline Audit and Evaluation*, (Heritage Council, 2007).

6   Quoted by Coones, P., in 'Geographical Approaches to the Study of Landscape', *Landscape Study and Management*, ed. F. H. A. Aalen (Boole Press, Dublin, 1996), p. 18.

7   Lennon, J., 'The Evolution of Landscape Conservation in Australia: Reflections on the relationship of nature and culture' *The Protected Landscape Approach: Linking Nature, Culture and Community*, eds. J. Brown, N. Mitchell, M. Beresford (IUCN, Switzerland & Cambridge, 2004).

8   Keller, Christian, 'The Theoretical Aspects of Landscape Study', *Decoding the Landscape*, ed. T. Collins (Centre for Landscape Studies, Galway, 1994), p. 91.

9   Fitzpatrick, E., 'Parley sites of Ó Néill and Ó Domhnaill in late sixteenth-century Ireland', *Regions and Rulers in Ireland, 1100– 1650*, ed. D. Edwards (Four Courts Press, Dublin, 2004), p. 201–216.

10 Tentative List and World Heritage Status – Research Document (Department of EHLG Wexford 2008), p. 15.

11 Phillips, A., 'Landscape as a meeting ground: Category V Protected Landscapes/Seascapes and World Heritage Cultural Landscapes' p. 6. *The Protected Landscape Approach: Linking Nature, Culture and Community*, eds, J. Brown, N. Mitchell, M. Beresford (IUCN, Switzerland & Cambridge, 2004).

12 Lennon, J., *op. cit.*, p. 6.

13 Aalen, F. H. A., *op. cit.*, p. 6.

CHAPTER 2

1   MS (RIA 23N II 178) Cormac mac Cuilennáin, King-Bishop of Cashel, who died in 908, quoted in Robinson T., *Stones of Aran: Labyrinth* (Lilliput Press, Dublin, 1995).

2   Feehan, J., 'The Geology of the Aran Islands', *The Book of Aran*, ed. J. Waddell, J. W. O'Connell, A. Korff (Tir Eolas, Galway, 1994), pp. 17–19.

3   Robinson, T., *Stones of Aran: Labyrinth* (Lilliput Press, 1995), p. 20.

4   Feehan, J., *op. cit.*, p. 18.

5   *Ibid.*, p. 19.

6   Daly, E. P., 'A Hydrogeological Investigation on Inishmaan, Aran Islands' (GSI Dublin, Internal Report, 1977).

7   Kinahan, G. H. & Foot, F. J., *6" field sheets, 1861* (GSI Archive).

8   Simms, M., *Exploring the Limestone Landscapes of the Burren and the Gort Lowlands* (Burrenkarst.com, Belfast, 2001), p. 11.

9   Feehan, J., *op. cit.*, p. 19.

10 Kinahan, G. H. & Foot, F. J., *6" field sheets, 1861* (GSI Archive).

11 Feehan, J., *op. cit.*, p. 20.

12 *Ibid.*, p. 29.

13 Waddell, J., 'The Archaeology of the Aran Islands', *The Book of Aran*, eds. J. Waddell, J. W. O'Connell, A. Korff (Tir Eolas, Galway, 1994), pp. 74–81.

14 Neolithic settlement is thought to have occurred near funerary monuments. See Aalen, F. H. A., *Man and the Landscape in Ireland* (Academic Press, 1978), p. 50.

15 Caulfield, S., 'The Neolithic Settlement of North Connaught', *Landscape Archaeology in Ireland*, ed. T. Reeves-Smyth, F. Hammond (B.A.R., Oxford, 1983), p. 213.

16 Aalen. F. H. A., *op. cit.*, p. 76.

17 *Ibid.*, p. 84.

18 Rynne, E., 'Dún Aengus – Fortress or Temple?' *An Aran Reader*, eds. B. and R. Ó hEithir (Lilliput Press, Dublin, 1999).

19 Aalen, F. H. A., *op. cit.*, p. 101.

20 Kelly, F., *Early Irish Farming: a study based mainly on the law-texts of the 7th and 8th centuries AD*, (Dublin Institute for Advanced Studies, 1998), pp. 404–406.

21 Kinahan, G. H., *op. cit.*, p. 34.

22 The notion of 'freehold' had a different meaning under the Gaelic system. See Nicholls, K., 'Land, Law and Society in 16th-Century Ireland' (O'Donnell Lecture, NUI, Dublin 1976).

23 O'Connell, J. W., 'History and the "Human Kingdom": A context for what follows', *The Book of Aran*, ed. J. Waddell, J. W. O'Connell, A. Korff (Tír Eolas, Galway, 1994), pp. 71–74.

24 Graham, J. M. 'Rural Society in Connaught 1600–1640', *Irish Geographical Studies*, eds. N. Stephens and R. E. Glasscock (Queen's University Belfast, 1970), and also McErlean, T., 'The Irish Townland System of Landscape Organisation', Landscape Archaeology in Ireland, eds. T. Reeves-Smyth, F. Hammond (B.A.R. Oxford, 1983). The early work of the geographer Jean Graham in regard to the land-division system in Connaught has been very informative and has influenced much of what follows on the subject, and also Thomas McErlean on the townland matrix.

25 Nicholls, K., 'Gaelic Society and Economy', *A New History of Ireland* Vol. 2, ed. Cosgrove (Oxford University Press, 1987), p. 407.

26 Kelly, F., *op. cit.*, p. 402.

27 Graham, J. M., *op. cit.*,

28 Graham, J. M., *op. cit.*, p. 193.

29 Nicholls, K., *op. cit.*, p. 408.

30 Jones Hughes, T., 'Town and Baile in Irish Place-Names', *Irish Geographical Studies*, eds. N. Stephens and R. E. Glasscock (Queen's University Belfast, 1970), p. 255.

31 Simms, A., 'Core and Periphery', *Common Ground: Essays on the Historical Geography of Ireland.*, eds. Smyth, Whelan (Cork University Press, 1988), pp. 34–38.

32 Andrews, J. H., *Plantation Acres*, (Ulster Historical Foundation, 1985), p. 32.

33 Mac Giolla Choille, B., ed. *Books of Survey and Distribution*, Vol. 3 (P.R.O.I., Dublin, 1962).

34 Petty, W., 'Political Anatomy' p.373 quoted in Andrews, J. H., *A Paper Landscape: The Ordnance Survey in Nineteenth-Century Ireland* (Four Courts Press, Dublin) Appendix A, The Spring Rice Report p. 301.

35 Andrews, J. H., *A Paper Landscape: The Ordnance Survey in Nineteenth-Century Ireland* (Four Courts Press Dublin, 2001), p. 14.

36 Ibid., Appendix A: Spring Rice Report, p. 304.

37 *Ibid.*, p. 22.

38 McErlean, T., *op. cit.*, p. 316.

39 McErlean, T., *op. cit.*, p. 335.

40 *Ibid.*, p. 334.

41 Freeman, M., *The Compossicion Booke of Conought* (transcribed and edited in 1936), copied in 1700 from original document of 1585, p. 52.

42 Books of Survey and Distribution, *op. cit.*

43 O'Flaherty, R., *A Chorographical Account of West or h-Iar Connaught*, written 1684 (ed. James Hardiman, Dublin, 1846; facsimile reprint, Galway, 1978), p. 76.

44 Robinson, T., *Oileáin Árann: a map of the Aran Islands Co. Galway* (Folding Landscapes 1996) The names of the quarters that follow this section are taken from this map.

45 O'Donovan J., *Ordnance Survey Letters County Galway* (printed volume RIA), p. 107.

46 Byrne, F. J., Irish Kings and High Kings (Four Courts Press, Dublin, 1973, 2001), p. 178.

47 See note 44.

48 My research led me to speculate that Gort na gCapall might contain a second *ceathrú* because of its large size and the amount of fertile land within it.

49 Kelly, F., *op. cit.*, pp. 363–364.

50 *Teampaill*, literally 'temple', refers to an early church. 'Bowen (1969) has described how the place-name elements "diseart" and "teampall" occur mainly in the south meaning "desert" and "temple" and are influenced by the eastern Mediterranean along the ancient Atlantic seaways and without the mediating influence of the Gallo-Roman world." Aalen, F. H. A., *Man and the Landscape in Ireland* (Academic Press, 1978), p. 103.

51 Robinson, T., *Oileáin Árann*, (text which accompanies map, T. D. Robinson, 1980), *op. cit.*

52 O'Donovan, J., *op. cit.*, p. 124.

53 Source: Dara Beag Ó Fátharta of Inis Meáin in conversation. He also said that this was really a *leath-ceathrú* (half quarter).

54 Andrews, J. H., *Plantation Acres* (Ulster Historical Foundation, 1985), p. 10.

55 Piers, H., A *Chorographical Description of the County of Westmeath*, first published in 'Collectanea de rebus Hibernicus,' Vol. 1 by C. Vallancey, 1786 (Meath Archaeological and Historical Society, Tara,1981), p. 116.

56 Source: NPWS Galway 2003. Both lakes are good examples of karstic lagoons. Loch an Charra in Cill Rónáin is a karstic saline lake lagoon with underground connections to the sea. An Loch Mór also has connections to the sea.

57 Dunford, B., *Farming and the Burren* (Teagasc, Dublin, 2002), p. 10.

58 Robinson, T., *Oileáin Árann: a map of the Aran Islands Co. Galway* (Folding Landscapes 1996).

59 The short-story writer Liam Ó Flaithearta, who was from Gort na gCapall, uses the word in his story, 'Teangabháil'. A young woman is riding a horse along the strand, watched by a group of men who are gathering seaweed. One man says, 'B'fhearr liom mac ona broinn ná cnagaire talún' (I would rather a son from her womb than a *fourth* of land) (translation L.O'F, my italics).

60 Burke, O. J., *The South Isles of Aran* (County Galway) (London, 1887), p. 69.

61 Source: Mairín Bean Uí Fhlaithbheartaigh, Gort na gCapall, in conversation, July 2004: while collecting the last census on Inis Meáin it was found that people there described their landholding in *cnagaire* and half cartrons and not in acres.

CHAPTER 3

1 Rynne, E., 'Dún Aengus – Fortress or Temple?' *An Aran Reader*, eds. B. and R. O hEithir, (Lilliput Press, Dublin, 1999), p. 261.

2 McCourt, D. 'The Dynamic Quality of Irish Rural Settlement', *Man and his Habitat: essays presented to Emyr Estyn Evans*. eds. R. Buchanan, E. Jones, D. McCourt, (Routledge & Kegan Paul, London, 1971), p. 153.

3 Whelan, K., 'Settlement Patterns in the West of Ireland in the Pre-Famine Period', *Decoding the Landscape*, ed. Collins, (SSRC Galway 1994), p. 69.

4 McErlean, T., 'The Irish Townland System of Landscape Organisation' *Landscape Archaeology in Ireland*, eds. T. Reeves-Smyth, F. Hammond (B.A.R. Oxford 1983), p. 355.

5 O'Flaherty, T., 'Coming Home', *An Aran Reader*, eds. B. and R. Ó hEithir, (Lilliput Press, Dublin, 1999), p. 180.

6 Robinson, T., *op. cit.*, p. 317.

7 Mná Fiontracha, *Árainn: Cosáin an tSaoil*, (Bailiúchán Bhéaloideas Árann, Árainn, 2003), p. 35.

8 Kelly, F., *Early Irish Farming: a study based mainly on the law-texts of the 7th and 8th centuries* AD, (Dublin Institute for Advanced Studies, 1998), p.370–374, 391.

9 Feehan, John, *Farming in Ireland: History, Heritage and Environment*, (UCD, Dublin, 2003), p. 87.

10 Source: Tomás Ó Fátharta, Eochaill, in conversation, summer 2004.

11 O'Connell, J. W., 'The Rediscovery of the Aran Islands in the 19th Century', *The Book of Aran*, eds. J. Waddell, J. W. O'Connell, A. Korff, (Tir Eolas, Galway, 1994), p. 194.

12 Robinson, T., *Stones of Aran: Labyrinth*, (Lilliput Press, Dublin, 1995), p. 40.

13 *Ibid.*, p. 379.

14 Andrews, J. H., A *Paper Landscape: The Ordnance Survey in Nineteenth-Century Ireland* (Four Courts Press, Dublin, 2001), p. 104.

15 *Ibid.*, p. 105.

16 Robinson, T., *op. cit.*, p. 382.

17 O'Flaherty, T., *op. cit.*, p. 178.

18 Powell, A., *Oileáin Árainn: Stair na n-Oileáin go dtí 1922* (Wolfhound, Dublin, 1984), pp. 69–76.

19 Robinson, T., *op. cit.*, p. 246.

20 Source: Dara Beag Ó Fátharta, Inis Meáin, September 2004, in conversation.

21 Robinson, T., *op. cit.*, p. 267. A story told by Seán Gillan of Oatquarter.

22 *Ibid.*, p. 247.

23 Feehan, J., 'The Geology of the Aran Islands', *The Book of Aran*, eds. J. Waddell, J. W. O'Connell, A. Korff, (Tir Eolas, Galway, 1994), p. 29.

24 O'Flaherty, J. T., A *Sketch of the History and Antiquities of the Southern isles of Aran, lying off the West coast of Ireland; with Observations on the Religion of the Celtic nations, Pagan Monuments of the Early Irish, Druidic Rites, &c* (Royal Irish Academy, Dublin, 1824), p. 56.

25 O'Flaherty, J. T., *op. cit.*, p. 58.

26 *Ibid.*, p. 57.

27 Powell, A., *op. cit.*, p. 58.

28 Drew, D., *Ancient Field Boundaries, Ballyelly–Coolmeen area, Slieve Elva,* (after Plunkett-Dillon) IQUA Field Guide No. 18.

29 I am grateful to John Feehan for this suggestion.

30 Aalen, F. H. A., *Man and the Landscape in Ireland* (Academic Press, London 1978), p. 59.

31 Waddell, J., 'The Archaeology of Aran', *The Book of Aran*, eds. J. Waddell, J. W. O'Connell, A. Korff, (Tír Eolas, Galway, 1994), p. 99.

32 O'Flaherty, T., *op. cit.*, p. 178.

33 Mac Giolla Choille, B., ed., *Books of Survey and Distribution*, Vol. 3 (1962), P.R.O.I. Dublin

34 Petty, W., *The History of the Survey of Ireland*, commonly called The Down Survey (The Irish Archaeological Society Dublin, 1851), p. 57.

35 Feehan, J., *Farming in Ireland; History, Heritage and Environment* (UCD Faculty of Agriculture, Dublin 2003), p. 60.

36 Feehan, J. *op. cit.*, p. 86. 'During the Revolutionary wars that swept across the country in the twelve years after 1641, the ruin of agriculture was a specific target. A letter sent to the Parliament of England by the Commonwealth Commissioners on 1st July 1650 described how Colonel Hewson had the week before marched into Wicklow and "Doth now intend to make use of scythes and sickles that were sent over in 1649, with which they intend to cut down the corn growing in those parts which the enemy is to live upon in the winter time, and thereby for want of bread and cattle the Tories may be destitute of provisions, and so forced to submit and quit those places".'

37 Piers, H., *op. cit.*, p. 116.

38 Powell, A., *op. cit.*, p. 69–76.

CHAPTER 4

1   Dunford, B., *Farming in the Burren* (Teagasc, Dublin 2002), p. 39.

2   Roden, C., 'The Aran Flora', *The Book of Aran*, edited by J. Waddell, J. W. O'Connell, A. Korff, (Tír Eolas, Galway, 1994), p.38.

3   Dunford, B., *op. cit.*, p. 13.

4   Robinson, T., *Notes to Oileáin Árann, Map of the Aran Islands* (T. D. Robinson, Galway, 1980).

5   Source: NPWS Galway 2003.

6   Synge, J. M., *The Aran Islands* (first published 1907) (Penguin, London, 1992), p. 32.

7   Source: in conversation with Dara Beag Ó Fátharta, Inis Meain. (b. *circa* 1925) and Tomás Ó Fátharta, Eochaill, Árainn (b.1922).

8   Source: Tomás Ó Fáthartha, Eochaill, Árainn, in conversation July 2004.

9   McAfee, P., *Irish Stone Walls*, (O'Brien Press, Dublin, 1997), p. 104.

10 Source: Dara Beag Ó Fátharta, Inis Meáin, September 2004.

11 Robinson, T., *op. cit.*, p. 268.

12 Robinson, T., *op. cit.*, p. 17.

13 Robinson, T., *op. cit.*, p. 329.

CHAPTER 5

1   Aalen, F. H. A., 'The Irish rural landscape: synthesis of habitat and history', *Atlas of the Irish Rural Landscape*, eds. F. H. A. Aalen, Kevin Whelan, Matthew Stout (Cork University Press, Cork, Ireland, 1997), p. 26.

2   Kelly, F., *Early Irish Farming: a study based mainly on the law-texts of the 7th and 8th centuries AD*, (Dublin Institute for Advanced Studies, 1998), p. 373.

3   Dunford, B., *Farming in the Burren*, (Teagasc, Dublin, 2002), p. 34.

4   Feehan, J., *Farming in Ireland: History, Heritage and Environment* (UCD, Dublin, 2003), p. 504.

5   Cole, L. & Phillips, A., 'Conserving Agro-biodiversity in England's Protected Landscapes', *Protected Landscapes and Agro-biodiversity Values*, eds. T. Amend, J. Brown, A. Kothari, A. Phillips, S. Stolton, (IUCN & GTZ Heidelberg 2008), p. 120.

6   Feehan, J., *op. cit.*, p. 503.

7   Bell, P., *Environmental Farming: A Guide to the Rural Environment Protection Scheme*, (Farrelly, Navan 1996), p. 2.

8   Feehan, J., *op. cit.*, p. 507.

9   Source: National Parks and Wildlife Service, Galway, 2003.

10 Source: Department of Agriculture, Galway, 2004.

11 *Ibid.*

12 Source: National Parks and Wildlife Service, Galway, 2003.

13 Bell, P., *op. cit.*, p. 2.

14 Scannell, Y., *The Habitats Directive in Ireland*, (Trinity College, Dublin 1999), p. 158.

15 Bell, P., *op. cit.*, p. 6.

16 Source: Department of Agriculture, Galway, 2004.

17 Scannell, Y., *op. cit.*, p. 161.

18 *Ibid.*, p. 161.

19 *Ibid.*, p.159.

20 Bohnsack U. & Carracan P., *An Assessment of Farming Prescriptions Under the Rural Environment Protection Scheme in the Uplands of the Burren Karstic Region. Co. Clare* (The Heritage Council, Dublin 1999) .

21 Bell, P., *op. cit.*, p. 15.

22 *Ibid.*, p. 7.

23 Conditions for the Conservation of the Burren to be applied under the Rural Environment Protection Scheme, 1995.

24 Bell, P., *op. cit.*, p. 82.

25 Source: Mike O'Flaherty, Teagasc Adviser Galway.

26 *Ibid.*

27 Source: Cyril Ó Flaithbheartaigh, Eochaill, in conversation.

28 Feehan, J. *op. cit.*, p. 505.

29 Dunford, B. *op. cit.*, p. 14.

30 Bohnsack U. & Carracan P. *op. cit.*, p. 57.

31 John O'Donohue, quoted in Dunford, B., *op. cit.*, p. 34.

32 Villalón, A., 'World Heritage inscription and challenges to the survival of community life in Philippine cultural landscapes', *The Protected Landscape Approach: Linking Nature, Culture and Community*, eds. J. Brown, N. Mitchell, M. Beresford (IUCN, Switzerland & Cambridge, 2004), p. 97.

33 Nozawa, C., Malingan, M., Plantilla, A. & Ong, J., 'Evolving culture, evolving landscapes: The Philippine rice terraces', *Protected Landscapes and Agro-biodiversity Values*, eds. T. Amend, J. Brown, A. Kothari, A. Phillips, S. Stolton, (IUCN & GTZ Heidelberg 2008), p. 71–93.

34 *Evaluation of Cultural Properties – Addendum*, World Heritage Convention, 28th Ordinary Session, ICOMOS 2004, p. 21–26.

35 State of Conservation of World Heritage Sites in Europe, Summary II (UNESCO 2006).

36 Lisitzin, K. & Stovel, H., 'Training Challenges in the Management of Heritage Territories and Landscapes', p. 34, *Cultural Landscapes: the Challenges of Conservation* (UNESCO Paris 2003).

37 Nozawa, C., Malingan, M., Plantilla, A. & Ong, J., *op. cit.*, p. 89.

38 Bohnsack, U. & Carracan, P., *op. cit.*, p. 61. 'Schemes of this nature have successfully been used on the continent, e.g. the "Pastured Steers from the Rhon scheme, the Pastured Heifers From St. Engmar" scheme or the "Rhon Lamb" scheme.'

39 Cole, L. & Phillips, A. *op. cit.*, p. 122.

40 *Ibid.*, p. 123.

41 *Ibid.*, p. 125.

42 *Ibid.*, p. 125.

Bibliography

MANUSCRIPTS AND MAPS

Trinity College Map Library:
Ordnance Survey First Edition Maps, surveyed 1839, printed 1841

Library of the School of Architecture, UCD:
Ordnance Survey Second Edition Maps, surveyed 1899, printed 1899

Geological Survey of Ireland Archive:
Geological Survey of Ireland (1867) 6" maps

Kinahan, G. H., Leonard, H., Cruise, R. J., *Memoirs of the Geological Survey, County of Galway* (Alexander Thoms, Dublin, 1871)

Daly, E. P., 'A Hydrogeological Investigation on Inishmaan, Aran Islands' (GSI Dublin, Internal Report, 1977)

Royal Irish Academy:
O'Donovan, John, *Ordnance Survey Letters County Galway* (printed volume)

O'Flaherty, J. T., A Sketch of the History and Antiquities of the Southern isles of Aran, lying off the West coast of Ireland; with Observations on the Religion of the Celtic nations, Pagan Monuments of the Early Irish, Druidic Rites, &c, (Royal Irish Academy, Dublin, 1824)

National Archive:
First Edition Ordnance Survey, Surveyors' notebooks
Books of Survey and Distribution, B. Mac Giolla Choille ed., vol. 3 (1962)

Library of University College Dublin:
The Compossicion Booke of Conought (transcribed and edited by A. M. Freeman, 1936), copied in 1700 from original document of 1585

The Annals of Connacht, edited and transcribed by A. M. Freeman (Dublin, Institute for Advanced Studies, 1944) Special Collections:

Burke, Oliver. J., *The South Isles of Aran* (County Galway) (London, 1887)

PUBLISHED WORKS

Aalen, F. H. A., *Man and the Landscape in Ireland*, (Academic Press, London,1978)

Aalen, F. H .A., ed. *The Future of the Irish Landscape*, (Trinity College, Dublin, 1985)

Aalen, F. H. A., ed. Landscape Study and Management, (Trinity College, Dublin,1996)

Aalen, F. H. A., 'The Irish Rural Landscape, synthesis of habitat and history', 'The Making of the Rural Landscape', *Atlas of the Irish Rural Landscape*, ed. F. H. A. Aalen, Kevin Whelan, Matthew Stout, (Cork University Press, Cork, Ireland, 1997)

Andrews, J. H., *Plantation Acres*, (Ulster Historical Foundation, 1985)

Andrews, J. H., *A Paper Landscape: The Ordnance Survey in Nineteenth Century Ireland*, (Four Courts Press, Dublin, 2001)

Bell, Patrick, *Environmental Farming: A Guide to the Rural Environment Protection Scheme*, (Farrelly, Navan, 1996)

Bohnsack U. & Carracan P., *An Assessment of Farming Prescriptions Under the Rural Environment Protection Scheme in the Uplands of the Burren Karstic Region, Co. Clare*, (The Heritage Council, Dublin, 1999)

Buchanan, R., 'Rural Settlement in Ireland', *Irish Geographical Studies: in honour of E. Estyn Evans*, ed. N. Stephens and R. E. Glasscock, (Queen's University, Belfast, 1970)

Byrne, Francis J., *Irish Kings and High Kings*, (Four Courts Press, Dublin (first edition 1973) 2001)

Caulfield, Seamus, 'The Neolithic Settlement of North Connaught', *Landscape Archaeology in Ireland*, ed. T. Reeves-Smyth, F. Hammond, (B.A.R. Oxford, 1983)

Cole, L. & Phillips, A., 'Conserving Agro-biodiversity in England's Protected Landscapes', *Protected Landscapes and Agro-biodiversity Values*, ed. T. Amend, J. Brown, A. Kothari, A. Phillips, S. Stolton, (IUCN & GTZ Heidelberg 2008)

Drew, David, 'The Burren County Clare', *Atlas of the Irish Rural Landscape*, ed. F. H. A. Aalen, Kevin Whelan, Matthew Stout (Cork University Press, Cork, Ireland, 1997)

Drew, D., 'Ancient Field Boundaries, Ballyelly-Coolmeen area, Slieve Elva', (after Plunkett-Dillon), *IQUA Field Guide No. 18*

Duffy, Patrick J., Edwards, David, Fitzpatrick, Elizabeth, ed., *Gaelic Ireland c.1250 – c.1650: Land, Lordship & Settlement*, (Four Courts Press, Dublin, 2001)

Dunford, Brendan, *Farming in the Burren*, (Teagasc, Dublin, 2002)

Expert Advisory Group, *Tentative List and World Heritage Status – Research Document* (Department of EHLG Wexford 2008)

Feehan, John, 'The Geology of the Aran Islands', *The Book of Aran*, ed. J. Waddell, J. W. O'Connell, A. Korff, (Tir Eolas, Galway, 1994)

Feehan, John, *Farming in Ireland: History, Heritage and Environment*, (UCD, Dublin, 2003)

Fielden, Bernard, Jokilehto, Jukka, *Management Guidelines for World Cultural Heritage Sites*, (ICCROM, Rome, 1998)

Fitzpatrick, E., 'Parley sites of Ó Néill and Ó Domhnaill in late sixteenth-century Ireland', *Regions and Rulers in Ireland, 1100–1650*, ed. D. Edwards, (Four Courts Press, Dublin, 2004)

Graham, Jean M., 'Rural Society in Connaught 1600–1640', *Irish Geographical Studies: in honour of E. Estyn Evans*, ed. N. Stephens and R. E. Glasscock, (Queen's University, Belfast, 1970)

Hughes, T. Jones, 'Town and Baile in Irish Place-Names', *Irish Geographical Studies: in honour of E. Estyn Evans*, ed. N. Stephens and R. E. Glasscock, (Queen's University, Belfast, 1970)

ICOMOS 2004, *Evaluation of Cultural Properties – Addendum* World Heritage Convention, 28th Ordinary Session,

Kelly, F., *Early Irish Farming: a Study Based Mainly on the Law-Texts of the 7th and 8th Centuries AD*, (Dublin Institute for Advanced Studies, 1998)

Keller, Christian, 'The Theoretical Aspects of Landscape Study', *Decoding the Landscape*, ed. Collins, (The Centre for Landscape Studies Social Sciences Research Centre, UCG, Galway, 1994)

Kolbert, C. F. and O'Brien, T., *Land Reform in Ireland*, (University of Cambridge Department of Land Economy, Cambridge, 1975)

Kostof, Spiro, *The City Shaped*, (Thames and Hudson, London, 1994)

Lennon, J., 'The Evolution of Landscape Conservation in Australia: Reflections on the Relationship of Nature and Culture', *The Protected Landscape Approach: Linking Nature, Culture and Community* ed. J. Brown, N. Mitchell, M. Beresford, (IUCN, Switzerland & Cambridge, 2004)

Lisitzin, K. & Stovel, H., 'Training Challenges in the Management of Heritage Territories and Landscapes', *Cultural Landscapes: the Challenges of Conservation*, (UNESCO Paris 2003)

McAfee, Patrick, *Irish Stone Walls: History, Building, Conservation*, (O'Brien Press, Dublin, 1997)

McCourt, Desmond, 'The Dynamic Quality of Irish Rural Settlement', *Man and his Habitat: Essays Presented to Emyr Estyn Evans*, ed. R. Buchanan, E. Jones, D. McCourt, (Routledge & Kegan Paul, London, 1971)

McErlean, Thomas, 'The Irish Townland System of Landscape Organisation', *Landscape Archaeology in Ireland*, ed. T. Reeves-Smyth, F. Hammond, (B.A.R., Oxford, 1983)

Messenger, John C., 'The Supernatural and the Esthetic', *An Aran Reader*, ed. Breandán and Ruarí O hEithir, (Lilliput Press, Dublin, 1991)

Mitchell, Frank, and Ryan, Michael, ed., *Reading the Irish Landscape*, (first published in Dublin 1986) (revised edition: Townhouse, Dublin, 2001)

Mná Fiontracha, Árainn: *Cosáin an tSaoil*, (Bailiúchán Bhéaloideas Árann, Árainn, 2003)

Nicholls, Kenneth, 'Gaelic Society and Economy', *A New History of Ireland Vol. 2*, ed. Cosgrove, (Oxford University Press, 1987)

Nozawa, C., Malingan, M., Plantilla, A., Ong, J., 'Evolving culture, evolving landscapes: The Philippine rice terraces' *Protected Landscapes and Agro-biodiversity Values*, ed. T. Amend, J. Brown, A. Kothari, A. Phillips, S. Stolton, (IUCN & GTZ Heidelberg 2008)

O'Connell, J. W., 'History and the "Human Kingdom" context for what follows', *The Book of Aran*, ed. John Waddell, J.W. O'Connell, Anne Korff, (Tir Eolas, Galway), 1994

O'Donovan, J., ed. *Annals of the Kindom of Ireland by the Four Masters*, (Dublin 1848, 1851, 1856)

O'Flaherty, Liam, *The Stories of Liam O'Flaherty*, ed. Vivian Mercier, (Devin-Adair Company, New York, 1956)

ÓFlaithearta, Liam, *Dúil*, (an Chéad Chló 1953) (Sáirséal Ó Marcaigh, Baile Átha Cliath, 1989)

O'Flaherty, Roderic *A Chorographical Description of West or h-Iar Connaught*, (written 1684, ed. James Hardiman), (Dublin 1846; facsimile reprint Galway 1978)

O'Flaherty, Tom, 'Coming Home' *An Aran Reader*, ed. Breandán and Ruarí O hEithir, (Lilliput Press, Dublin, 1991)

Petty, William, *The History of the Survey of Ireland, Commonly Called The Down Survey*, (The Irish Archaeological Society Dublin, 1851)

Phillips, A., 'Landscape as a meeting ground: Category V Protected Landscapes/Seascapes and World Heritage Cultural Landscapes', *The Protected Landscape Approach: Linking Nature, Culture and Community*, ed. J. Brown, N. Mitchell, M. Beresford, (IUCN, Switzerland & Cambridge, 2004)

Piers, Sir Henry, *A Chorographical Description of the County of Westmeath*, first published in 'Collectanea de rebus Hibernicus,' vol. 1 by C. Vallancey, 1786, (Reprinted facsimile by Meath Archaeological and Historical Society, Skryne Castle, Tara, County Meath, 1981)

Powell, Antoine, *Oileáin Árann, Stair na n-oileán anuas go dtí 1922*, (Wolfhound Press, Dublin, 1984)

Scannell, Yvonne, *The Habitats Directive in Ireland*, (Trinity College, Dublin, 1999)

Robinson, Tim, *Oileáin Árann*, Map of the Aran Islands, Co. Galway (with introductory text) (T. D. Robinson, Galway, 1980)

Robinson, Tim, *Stones of Aran: Pilgrimage*, (Lilliput Press, Dublin, 1986)

Robinson, Tim, *Stones of Aran: Labyrinth*, (Lilliput Press, Dublin,1995)

Rynne, Etienne, 'Dún Aengus – Fortress or Temple?' *An Aran Reader*, ed. Breandán and Ruarí Ó hEithir, (Lilliput Press, Dublin, 1991)

Simms, Anngret, 'Core and Periphery', *Common Ground: Essays on the Historical Geography of Ireland*, ed. Smyth, Whelan, (Cork University Press, 1988)

Simms, M., *Exploring the Limestone Landscapes of the Burren and the Gort Lowlands*, (Burrenkarst.com, Belfast, 2001)

Synge, J. M., *The Aran Islands*, (London and Dublin 1907), republished with introduction and notes by Tim Robinson 1992, (Penguin Books, London, 1992)

Tratman, E. K. ed., *The Caves of North-West Clare, Ireland* by The University of Bristol Spelaeological society, (David & Charles, Newton Abbot, Devon, 1969)

Trudgill, Stephen, *Limestone Geomorphology*, ed. Clayton, K .M. (Longman, New York, 1985)

UNESCO 2006, *State of Conservation of World Heritage Sites in Europe, Summary II*

Villalón, A., 'World Heritage inscription and challenges to the survival of community life in Philippine cultural landscapes', *The Protected Landscape Approach: Linking Nature, Culture and Community*, ed. J. Brown, N. Mitchell, M. Beresford, (IUCN, Switzerland & Cambridge, 2004)

Waddell, John, 'The Archaeology of the Aran Islands', *The Book of Aran*, ed. John Waddell, J. W. O'Connell, Anne Korff, (Tir Eolas, Galway, 1994)

Whelan, K., 'Settlement Patterns in the West of Ireland in the Pre-Famine Period', *Decoding the Landscape*, ed. Collins, (The Centre for Landscape Studies Social Sciences Research Centre, UCG, Galway, 1994)

List of Figures

175

Glossary

Bronze Age   the period of human culture characterised by the use of bronze, in Ireland considered to have been between 2000 BC and 500 BC

cuesta   a hill or ridge with a steep face on one side and a gentle slope on the other

clachan   a group of houses laid out informally, forming a village

clochan   a drystone hut, usually circular in plan, with corbelled roof dating from the Middle Ages or earlier

erratic   stones or boulders, sometimes quite large, left behind by the retreating ice at the end of the last Ice Age 10,000 to 14,000 years ago

gavelkind   Anglo-Saxon term also known as partible inheritance; practice of land tenure that divided property equally among a family of heirs on the death of the father, in use in Ireland under the Brehon Laws and continued, although outlawed, until the nineteenth century

gryke   a deep fissure or cleft in a limestone surface

Iron Age   period of human culture when iron began to replace bronze as the predominant metal used, generally considered to have occurred in Ireland between 500 BC and AD 400

karren   minor forms of karst formed by solution of the surface of the stone

karst   an irregular limestone region with sinks, underground streams and caverns formed by water erosion

Neolithic   relating to the Stone Age period, characterised by the use of polished stone implements, farming and megalithic structures beginning in Ireland about 4000 BC until the Bronze Age (2000 BC)

orthogonal   intersecting or lying at right angles

partible   divided into parts

rendzina   belonging to a group of dark greyish-brown soils developed in high to moderate humidity, including shallow soils over limestone in which the topsoil is calcareous

rundale   land-division system which involved individually farmed plots around a hamlet or village (infield) and collectively farmed larger areas at a distance from the settlement (outfield), in use in Ireland in the seventeenth and eighteenth centuries and probably earlier

Index